Urban Sustainability

The Urban Sustainability Book Series is a valuable resource for sustainability and urban-related education and research. It offers an inter-disciplinary platform covering all four areas of practice, policy, education, research, and their nexus. The publications in this series are related to critical areas of sustainability, urban studies, planning, and urban geography.

This book series aims to put together cutting-edge research findings linked to the overarching field of urban sustainability. The scope and nature of the topic are broad and interdisciplinary and bring together various associated disciplines from sustainable development, environmental sciences, urbanism, etc. With many advanced research findings in the field, there is a need to put together various discussions and contributions on specific sustainability fields, covering a good range of topics on sustainable development, sustainable urbanism, and urban sustainability. Despite the broad range of issues, we note the importance of practical and policy-oriented directions, extending the literature and directions and pathways towards achieving urban sustainability.

The series will appeal to urbanists, geographers, planners, engineers, architects, governmental authorities, policymakers, researchers of all levels, and to all of those interested in a wide-ranging overview of urban sustainability and its associated fields. The series includes monographs and edited volumes, covering a range of topics under the urban sustainability topic, which can also be used for teaching materials.

Conghui Zhou

Principles and Methods of Pocket Park Planning and Design

东南大学出版社
SOUTHEAST UNIVERSITY PRESS

Springer

Conghui Zhou
School of Architecture
Southeast University
Nanjing, Jiangsu, China

ISSN 2731-6483 ISSN 2731-6491 (electronic)
Urban Sustainability
ISBN 978-981-96-5800-8 ISBN 978-981-96-5801-5 (eBook)
https://doi.org/10.1007/978-981-96-5801-5

Jointly published with Southeast University Press
The printed edition is not for sale in China Mainland.
Customers from China Mainland please order the print book from Southeast University Press.
ISBN of the Co-Publisher's edition: 978-7-5766-0260-9

National Natural Science Foundation of China—State Grid Corporation Joint Fund for Smart Grid, Science and Technology Projects of Ministry of Housing and Urban-Rural Development of China, 2018-K2-005

Translation from the Chinese Simplified language edition: "《口袋公园规划设计原理与方法》" by Conghui Zhou, © Southeast University Press 2022. Published by Southeast University Press. All Rights Reserved.

This Springer imprint is published by the registered company Springer Nature Singapore Pte Ltd.
The registered company address is: 152 Beach Road, #21-01/04 Gateway East, Singapore 189721, Singapore

Foreword

Next Generation Urban Infrastructure and Green Space Distribution

Although this extraordinary research is embedded in the particular context of Nanjing, the implications are global; not simply for urban green space distribution, but for a range of micro-infrastructural considerations. This study addresses the proliferation of micro-green public space development in Nanjing in the past decade. It also presents a major contribution to understanding the changing characteristics of urban infrastructure for cities everywhere. It challenges the normative limits of research in landscape architecture and urbanism. It expands the traditional definitions of urban infrastructure to include the micro-scale. It redefines the concept of urban nature as public good. It pushes the limits of public space planning theory, through a combination of development metrics, planning principles, and test bed verification.

The history of city-making is replete with strategies and rationales for incorporation of "urban green" space within urban development. In Europe and the United States, since the onset of the industrial revolution, large-scale urban green space has been considered an indispensable antidote to increasing densities and heightened industrial hazards. The most extreme historical response has been to de-densify cities; to incorporate nature within "sub-urban" aggregations under the rubric of "suburb" or "garden city." By contrast, more recently new high-rise building technologies has allowed nature to mitigate hyper-densities. Extreme green could be interspersed with inexpensive tall buildings serviced by elevators. Entire new cities could be reduced to a "park" inhabited by towers distributed on a green tableau as the "city-in the-park," rather than the traditional "park-in-the-city." This study confirms the efficacy of revisiting the "park-in-the-city" as the normative means for integration of green space within high urban densities.

Historically, the ideal of the "city-in-the-park" became widespread globally, in large part as response to health considerations. From a medical perspective, prominent in the advocacy of "urban green" strategies has been purposing nature as an antidote to disease; and such was especially important during the Tuberculosis pandemic of the 19th and 20th centuries, when the cure was exposure to "sun, space and green." Tubercular patients were relocated from the city to the countryside predominantly via long-term stays in rural sanitoria. Subsequent medical breakthroughs would change this formulation, including penicillin. What remained urbanistically, however, was the ideal of the tower-in-the-park, that can accommodate urban density within vast continuous green space. The city would be subservient to a park context, which has become prevalent in many cultural contexts as a convenient way to address both economic and political (ideological) realities. It is worth noting that the COVID pandemic of today has reinforced this ideal of escape to nature from cities, at least for those who can afford it.

Food production has also been historically prominent in the advocacy of urban green. For example, urban agriculture has been an important military consideration, especially for European walled cities of the 14th–16th centuries, when military sieges could last for prolonged periods requiring food production within the walls. And even in modern warfare, the "victory garden" was a requisite food source in both World War I and II in Europe and the United States. Apart food, mini-gardens contributed a psychological dimension to understanding the role of nature in nurturing well-being. The role of nature in the urban psyche is a subject that has been anecdotal for many decades, including today's discourse on climate adaptation, when the social science comes to the fore in strategizing about limiting our carbon-based lifeworlds and the realities of coastal withdrawal. Such is a discourse on infrastructure as "social condenser" and the reinvention of urban "production" in general within our transitions in thought about the relationship of urban density and urban context.

Historical rationales for spatial disposition of urban green space are being reconsidered, or at least the mandates are softening. The strategy of fine-grain distributed parks integrated within city fabric is gaining credibility for numerous reasons, with micro-green space distribution connected to other innovation in distributed infrastructure. The distribution question is relevant for many infrastructural applications including energy and water systems. The basic operation of urban infrastructure is evolving from the historic "modern" city dominated by "top-down" geometries—large-scale gridding that can be infilled incrementally, but that are overly deterministic in terms of formal outcomes. Now top-down is mitigated by the strategy of "bottom-up" distribution. There are diverse advantages to "leap-frogging" top-down that are crucial to developing urban economies where top-down is not even remotely possible. There are also advantages within so-called "advanced" urban economies where the social and spatial intransigence of top-down must be remediated.

For urban green, "bottom-up" incremental infill has become strategic, but with management challenges. The technology to manage new forms of infrastructural distribution is evolving, especially related to digital media. Social media is revolutionizing how urban infrastructure is deployed and managed. At the most basic level, real-time usage of public green space can be documented, through geo-located data capabilities. For the effective functioning of distributed infrastructure, this capability is essential. For example, in recent years there has been increasing use of social media sentiment data for urban green space research. Sentiment data can add significantly to research into the psychological benefits of urban parks, permitting fine-grain resourcing for mini-park planning strategies. For the maintenance of fine-grain park infrastructure, a level of monitoring is required that may effectively (and realistically) be provided through social media.

For monitoring and understanding the new infrastructural forms very important is the deployment of "hard" science with "design science" using a new level of research and development means that have no real precedent within design culture. With this new reality, the discipline of "landscape architecture" is evolving, such that research into form-based models beyond vegetation patterning is a crucial planning component. What better test bed for innovation than China? For the promise of next generation urban infrastructure, Chinese cities are of particular interest, in that densities are much higher than in cities in much of the rest of the world; combined with rather particular green space characteristics including a high proportion of public land combined with public space planning tools that may be more robust than in many other urban contexts. The breadth of this study provides good guidance for optimizing these advantages - and much more. This study deserves our careful consideration, as pioneering work in the development of next generation urban infrastructure everywhere.

Professor Richard Plunz
Columbia University
New York, USA

Professor Richard Plunz is Professor of Architecture at Columbia University. He has taught at Columbia since 1973, and served as Chair of the Division of Architecture. From 1992 to 2015, he also served as director of the post-professional Urban Design Program.

Acknowledgements I would like to express my gratitude to those who have made contributions at each phase of the publication of this book. First of all, I should thank the students and my team members from Southeast University who participated in the National Natural Science Foundation Project *"Research on the collaborative mechanism and key regulatory indicators of mini-park groups in high-density cities"*. Their work is the source of this book writing. The relevant work begins with the field survey and service analysis of individual pocket parks, in which Leshan Fu, Yanwen Xue, Zhijie Wang, Qinsi Zhao, Meng Xie, Mengjie Cui, and Ziyue Liu made great efforts, especially on the aspects of the original data collection, analysis, and validation. On this base, Jin Zhao, Yihuan An, Leshan Fu, Yanwen Xue, Shining Zhang, Bowen Liu, and Yuting Yan helped to expand the study scope from individual park to park group, specially focusing on the explorations of the service mechanism, structure features, and construction method of pocket park groups. Subsequently, the work of Liuyi Yang, Zhuoyang Jiang, Xinyu Wu, Luanxin Song, Yi Sun, and Xuan Zhu successfully introduced AI technologies into our study, by which many new study possibilities and application opportunities were created.

Second, I would like to thank our main collaborators Yu Zhang, Yi Wu, Xingtong Zhu, and Xiaozhao Liu from Jiangsu Institute of Urban Planning and Design for their tremendous support and input to our collaborative planning project "Yancheng Urban Park Sector Planning (2019–2030)" and the Science and Technology Plan of the Ministry of Housing and Urban Rural Development of China, which greatly contributed to the validation and verification of our research results.

Third, I must express my great gratitude to professor Richard Plunz of Columbia University, who generously wrote the foreword of this book and gave us numerous enlightenments and suggestions on this study. Furthermore, Professor Yuning Cheng of Southeast University and Professor Yunfeng Jin of Tongji University also kindly provide help and support to this study, on which I appreciate so much. Additionally, several renowned experts from the academic journals, universities, and planning and design institutes, including Professor Hexian Jin, Zhitao Sun, Jun Cheng, Lingchen Yang, Dou Dou, Jijiang Ding, Professor Zhen Xu, Professor Zheng Chen, Professor Yujia Zhai, Professor Yilun Xu, and Professor Xiaoqin Yan, also kindly provided precious comments and suggestions for us to improve this study.

Moreover, to make this book more readable for non-professional people, numerous figures and photos were added as an essential supplement to the texts. Excluding the figures and photos with source indications, the figures and photos in the book were drawn or photographed by our team members: Yihuan An, Jin Zhao, Meng Xie, Bowen Liu, Shining Zhang, Jiangyan Chen, Qinsi Zhao, Leshan Fu, Yanwen Xue, Zhijie Wang, Mengjie Cui, and Yan Yuting. In addition, Jin Zhao and Yihuan An also completed the most work of type setting, while Shining Zhang designed the cover. As the original draft of this book was written in Chinese which necessitates a holistic translation and multiple rounds of proof edit, this heavy work was jointly undertaken by Shining Zhang, Liuyi Yang, and Luanxin Song. Last but not least, I need to thank our executive editor Zhengxia Zhu of domestic side and the executive editors from Springer, as this book can never be successfully published without their sustained support.

Furthermore, fund support is also indispensable for our work. In the whole process of this study and publication, we got multiple fund supports including the National Natural Science Foundation of China (Grant No. 51978146), Science and Technology Projects of Ministry of Housing and Urban-Rural Development of China (Grant No. 2018-K2-005), and Excellent Textbook Construction Fund of Southeast University. We would like to thank the fund committees for their trust and endorsement of this study.

Competing Interests The author has no competing interests to declare that are relevant to the content of this manuscript.

Introduction

Today, urban green spaces are being recognized as a strategic asset of modern cities. As the development focus of more and more cities shifts from external expansion to internal renewal, the paradigm and evaluation standards of urban green space planning are being reformed. In this process, "existing lands" have replaced "incremental lands" as the main source of urban green space growth. For the high-density cities, the compact and intensive construction of the built-up areas directly leads to the miniaturization, decentralization, and fragmentation of the existing lands that are available for new green space installation. This severely limits the development feasibilities of large-sized green spaces, and on the other hand, brings numerous opportunities for the development of micro green spaces.

As the main form of micro green spaces, "pocket park" is a concept sharing high popularity, which is becoming the mainstay for improving the service and spatial structure of existing urban green space system in built environment. Taking Nanjing, China as an example, the development of pocket parks directly led to 10% increase of the per capita park area from 2013 to 2017. In Shanghai's 14th Five-Year Plan (2021–2025), the construction of pocket parks will contribute to more than half of the increment of city's outdoor recreational opportunities. As to our realistic lives, pocket parks not only provide new recreational sites for improving people's health and well-being, but also contribute to reshaping the modern lives by their flexible and innovative usage patterns. It is precisely these places as well as the diverse activities occurred within them that bring about the precious warmth, colour, and intimacy to urban spaces, which turns to be a unique symbol of the humanistic charm of modern cities.

The conceptual prototype of pocket parks can be traced back to the tiny green spaces developed from the bomb-damaged sites during the reconstruction of European cities after World War II. Although it was an expedient under the severe constraints of capital, labour, and materials, it achieved a success that was far beyond expectations, which also generated profound influences on the subsequent development of public spaces within US cities. In the 1960s and 1970s, the relevant construction practices of pocket parks in the urban renewal of US cities like New

York, Philadelphia, and Baltimore also endowed the concept of pocket parks with a global reputation.

However, when reviewing the development history of pocket parks, we found that although European and American cities accumulated massive experience on pocket park development, the systematic theories and methodologies that can support the pocket park development in a large scale scarcely appeared. Accordingly, it leads to a significant randomness in the spatial distribution and service setting of pocket parks, as most of them were spontaneously constructed and maintained by local residents and non-governmental organizations. This loose and random development mode can increase the size of local green spaces and effectively respond to the demands of local residents. However, due to the lack of systematic consideration and coordination in the large scale, the value of pocket parks is greatly constrained.

Compared to European and American cities, the density of Chinese cities is much higher while most cities' green space development seriously lagged behind the growth of urban population in the urbanization. It directly leads to the extreme scarcity of green spaces in numerous built-up areas such as old cities, historic zones, and CBDs and generates enormous development demands of pocket parks. Meanwhile, China's land ownership and management system also provides much more feasibilities for developing pocket parks in a systematic mode. In order to fully unleash the potential value of pocket parks at city scale, it necessitates the construction of a holistic coordination and control system at both theoretical and methodological dimensions.

Our team paid attention to the planning and design of pocket parks in the early 2010s when we noted the shift of planning paradigm in urban renewal. Subsequently, with the support of the National Natural Science Foundation and the Science and Technology Plan of the Ministry of Housing and Urban Rural Development of China, we conducted research on urban green spaces' adaptive mechanism in high-density environments, fine classification framework, and intensive spatial distributional technology, through which we found that in urban renewal process, the granularity of green space planning and design objects is being increasingly refined, and pocket parks have played the dominant role in the sustained provision of new urban green spaces.

However, we also noted that the forms of pocket parks both includes "formal green spaces", occupying complete and independent land parcels in the land use category of "green spaces", and "informal green spaces" that are attached to the land parcels of other uses, which greatly complicates pocket parks' planning and management. Furthermore, significant differences were observed in the service and spatial attributes of pocket parks compared to conventional large-sized parks, resulting in the inapplicability of the classical theories and methods that were concluded from the large-sized park planning and design practices. Due to the lack of a systematic clarification of the conceptual connation, unique attributes, and planning and design principles of pocket parks, problems such as unclear concept definition, ambiguous service division, and inequitable spatial distribution frequently arose in the planning and design practices regarding to pocket parks.

In addition to the review of the relevant theories and practices, the observations, experiences, and perceptions from daily lives also provide a wealth of valuable ideas and materials for this book writing. In the old city of Nanjing where we have been working and living, there is a strong contrast between the high population density and scarce conventional parks (e.g. urban parks and community parks), forcing residents to use the pocket spaces, such as street corners, streetside spaces, and building gaps, for various daily recreational activities. Through a long-term observation and comparison of the usage characteristics of such spaces, we found that although they are fragmentarily-distributed and tiny-sized, they showed high effectiveness in response to the diverse demands of residents, sometimes even more effective than the larger parks.

The objective review and subjective experience jointly prompted us to rethink the appropriate service mode of green spaces within built-up areas and raised up two main questions: how can the service effectiveness of individual pocket park be guaranteed through planning and design, and how can the site selection and service division of multiple pocket parks be coordinated to maximize their potential value, by which the supply-demand gap of recreational resources in densely populated cities can be greatly filled? With these two questions in mind, we conducted a series of qualitative and quantitative studies on pocket parks' basic attributes, influencing factors of their service, collaborative mechanism, site selection methods and technologies, and institutional guarantee policies. The adopted methods include both traditional tools, such as questionnaire, on-site interview, and fixed-point shooting, and newly-emerged tools, such as big data and artificial intelligence (AI) technologies.

Finally, building on the relevant work of existing studies and practices, we classified and summarized our study results and findings and wrote this book. The book's content was divided into eight chapters: Chap. 1 introduced the concept origin, development background, value, and challenges of pocket parks; Chaps. 2 and 3 analysed the attributes and service mechanisms of pocket parks, respectively, focusing on their uniqueness compared to conventional large-sized green spaces; Chaps. 4 and 5 centred on the group construction, site selection, and service provision of pocket parks, on which "Park Group" was set as an independent chapter as we thought that specific highlights should be given to promote collaboration of pocket park group in order to maximize the system's effectiveness; Chap. 6 focused on the institutional guarantee in terms of planning system, control approaches, support policies, and implementation guarantee; Chaps. 7 and 8 provided two case studies focusing on the application of the planning methods and AI-based site selection technology, respectively.

In our opinion, this book can serve as a textbook for the undergraduate and graduate students in the disciplines of landscape architecture and urban planning, as well as a reference book for the professionals. Additional contributions of this publication are also expected to provide new ideas for promoting the paradigm shift of green space planning and design under urban renewal background. Furthermore, we attempted to use the language that is easy to understand and accompanied by a large number of visual illustrations in the writing of this book, through which we hope that the publication of this book would not only facilitate the professional teaching and practices, but also motivate more interests, attentions, and deliberations of the public towards urban green space planning and design, by which an active and effective public participation in the process of green space planning, design, construction, and maintenance can be achieved.

In our research process, we also deeply recognized that our current work is far from enough and perfect. For example, for the service mechanism of pocket parks, although our study has identified partial influencing factors and their modes of action, there are still many uncertain factors, especially on the non-physical aspects, calling for further exploration and identification to improve current planning and design guidelines. Meanwhile, compared to large-sized parks, we found that the interaction among pocket parks within a group is more complex, which might generate both competitive and collaborative effects. As this interaction effect is highly sensitive to the variation of park attributes and external environment, the current methods adopted in our studies cannot precisely trace it, which impedes the formulation of a refined guidance for constructing pocket park groups. It also requires for further investigation and verification based on more advanced tools of data collection and analysis. In terms of pocket parks' site selection and service provision, we mainly considered the demand disparities caused by different age groups. However, differences in residents' occupation, ethnic, gender, education, and other social attributes might also result in different recreational habits and demand. How to integrate the demand differences led by the various social attributes of user groups into planning and design consideration and further improve the accuracy of demand response, is also a topic worthy of further exploration. At the policy level, as the regularization and systematization of pocket park development are still in its early stages, there is still a long way to go in exploring a set of pocket parks' planning model, support system, and implementation mechanism that are suitable for the economic, cultural, and social characteristics of Chinese cities as well as the other cities worldwide. Therefore, we sincerely request any comments, suggestions, and corrections.

Contents

Chapter 1
Introduction

In modern cities, pocket parks are a type of non-conventional green spaces that is being popularized in recent decades. As the functional and spatial evolution of urban green space system is tightly associated with the conversion of urban function and morphology, pocket parks should not be separately seen as a randomly-emerged object in urban context. Instead, their emergence and popularization are more inclined to be regarded as a symbol of the transition of green space development, which is jointly triggered by multiple factors related to urban economic, society, and environment in a particular development phase of modern cities. Consequently, to better understand the concept of pocket parks, it is necessary to thoroughly review the pocket parks' origin, connotation, and value in the broad context of urban development process.

1.1 Urban Green Space Planning and Pocket Park

1.1.1 Space Formation of Modern Cities

The green space system of modern cities was initially formed in the First Industrial Revolution [1], when the urban space was restructured to adapt to cities' modern functions. In this process, although urban green spaces are seemed as functionally indispensable for modern cities, they need to compete with other land use categories, such as residential, industrial, and commercial lands, for land resource in development. However, as urban green spaces cannot directly generate economic returns, it is difficult for them to obtain the equivalent development opportunities as other land use categories through pure market-oriented mechanisms, requiring for the external intervention to compensate the deficiency in land resource distribution solely oriented by the market. To achieve this goal, early industrialized countries, such as United Kingdom and France, provided the earliest urban green spaces for

C. Zhou, *Principles and Methods of Pocket Park Planning and Design*, Urban Sustainability, https://doi.org/10.1007/978-981-96-5801-5_1

modern cities mainly through the publicization of royal gardens [2], conservation of public lands [3], and construction of new parks [2].

At this phase, most explorations on the ideal models of urban green space system were integrated into the classic theory and models of modern cities, such as the radiative green space system proposed by Ebenezer Howard in the model of "Garden City" [4] [Fig. 1.1], the broad matrix-axis green space system proposed by Le Corbusier in the model of "The Radiant City" [5] [Fig. 1.2], and the park-parkway system proposed by Frederick Law Olmsted based on the American-styled "Grid-Planned City" [6] [Fig. 1.3]. These models provided a theoretical basis for the early construction of urban green space system. In this period, urban planners and decision-makers were keen on using large-sized green spaces to dominate the formation of urban structure, such as the construction of the famous Birkenhead Park of Liverpool, Hyde Park of London, Central Park of New York. Considering the essential role of green spaces in urban development, the baseline land use quota or ratio was frequently used in the planning to guarantee the equitable status of green spaces in cities' land resource distribution (Table 1.1).

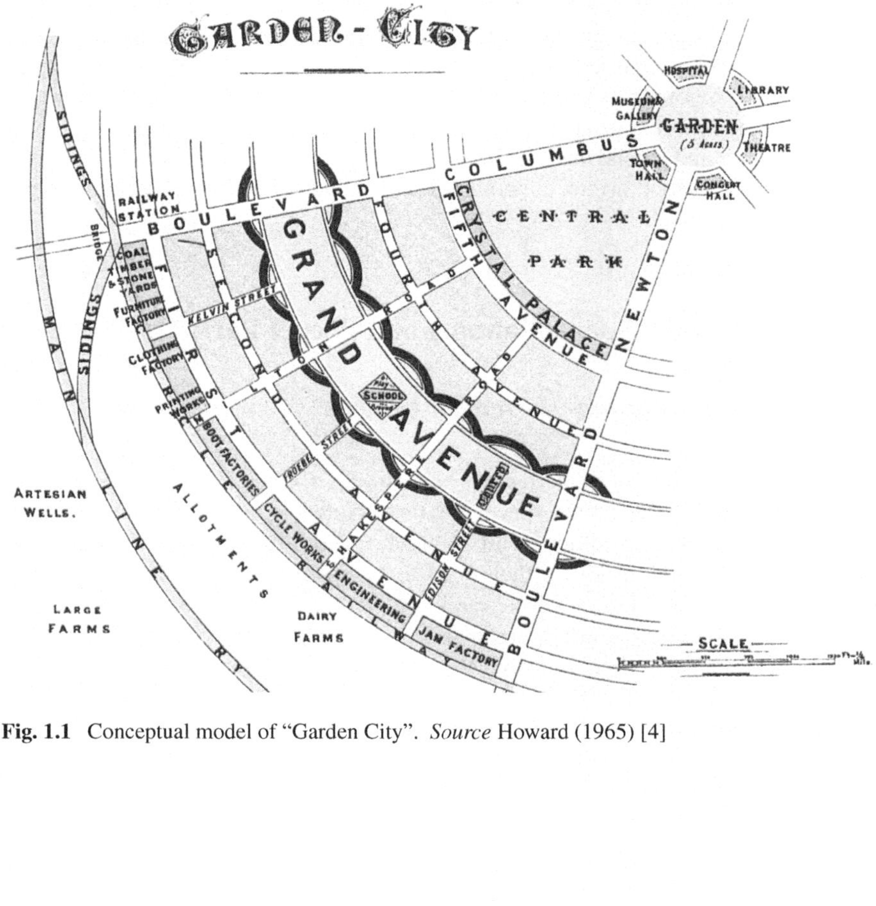

Fig. 1.1 Conceptual model of "Garden City". *Source* Howard (1965) [4]

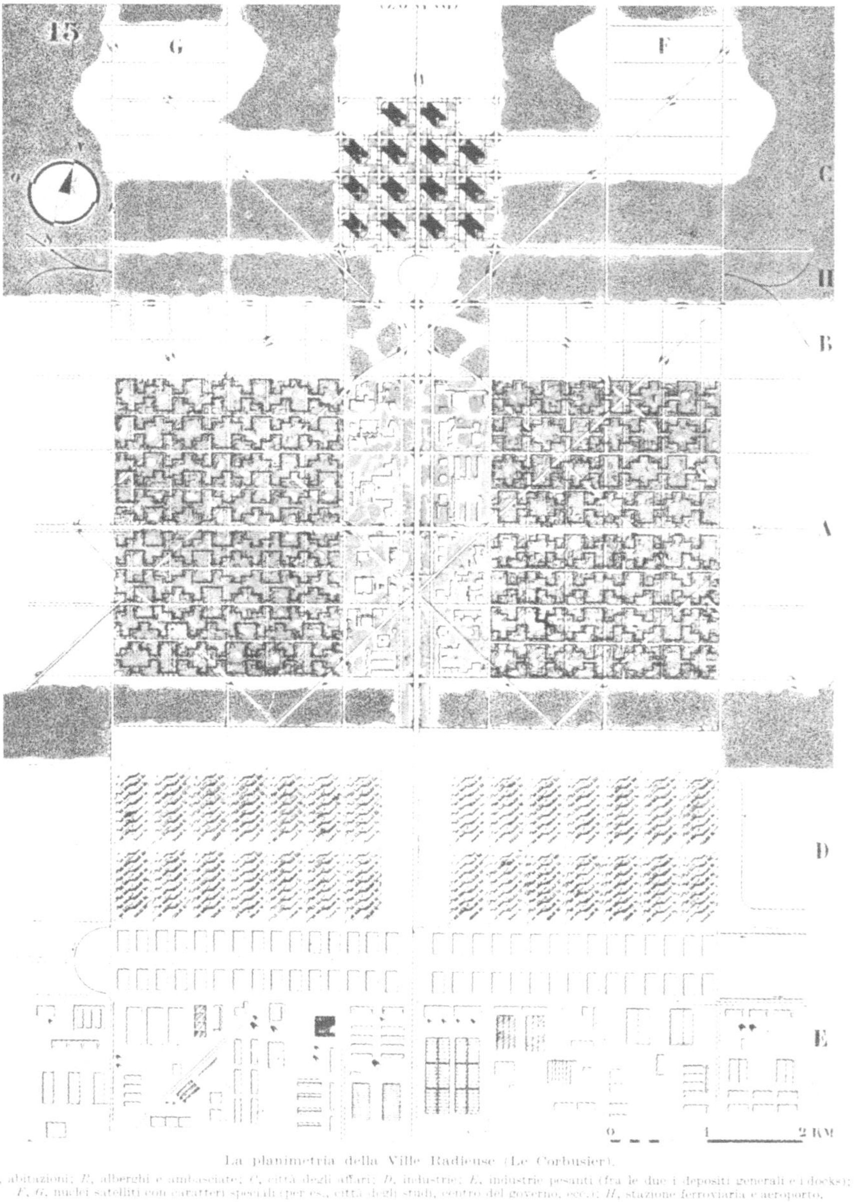

Fig. 1.2 Conceptual model of "The Radiant City". *Source* Corbusier (1964) [5]

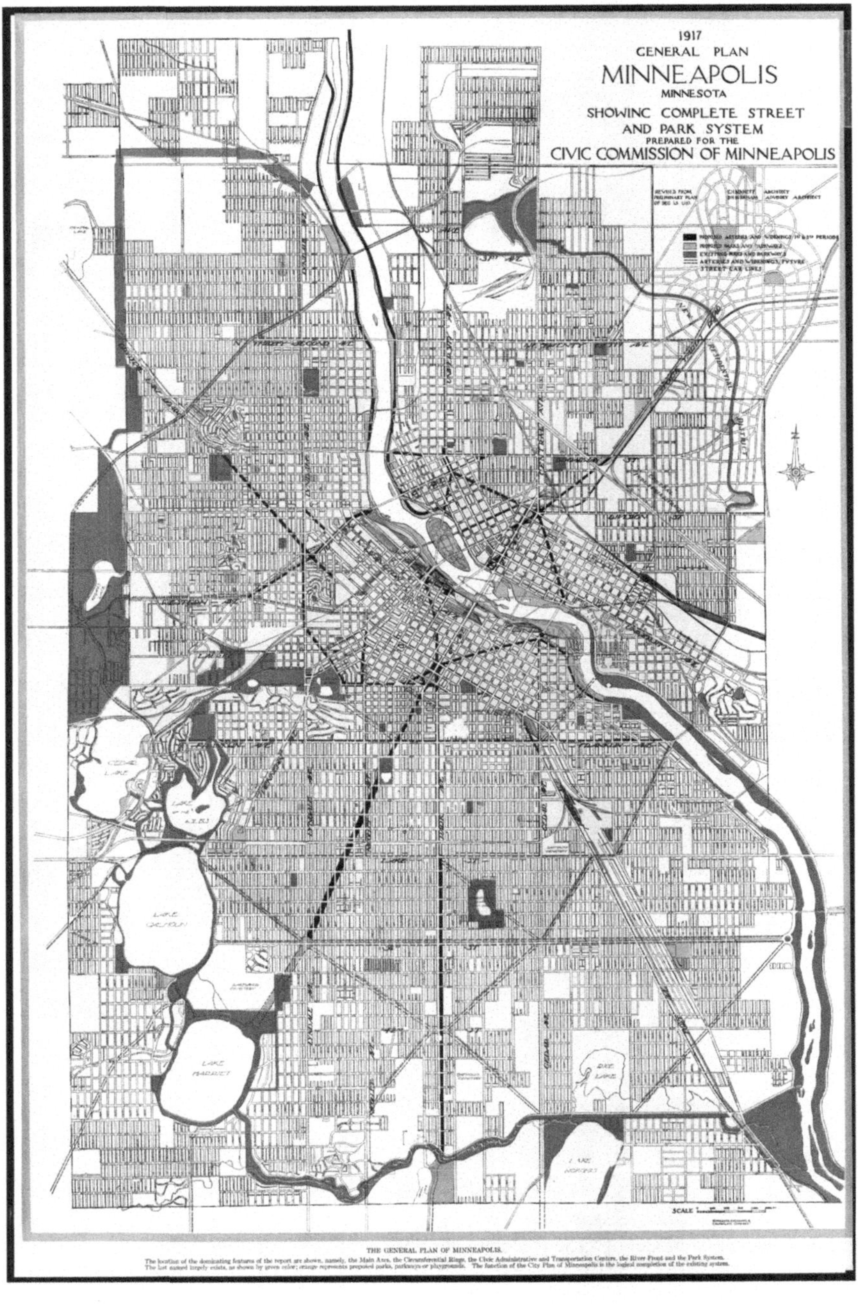

Fig. 1.3 Park system in the Minneapolis General Plan of 1917, USA. *Source* https://digitalcollections.hclib.org/digital/collection/p17208coll17/id/20, Accessed 15 April 2022

Table 1.1 Evolution process of urban green space planning

Urban development phases	Main goals in urban green space planning	Main approaches	Main objects	Key indicators
Formation of modern cities	Development opportunities guarantee	Construction of urban green space system	Large-sized green spaces	Baseline land use quota
Rise of residential planning	Spatial accessibility guarantee	Construction of urban green space hierarchical system	Green spaces of various levels	Accessibility indicators
Extensive renewal	Social equity guarantee	Optimization of entire system	Medium- and small-sized green spaces	Equity indicators
Micro-renewal	Demand response guarantee	Optimization of partial system	Micro-green spaces	Demand response indicators

1.1.2 Rise of Residential Planning

In the early urbanization, the centralization mode of urban development led to a series of environmental and traffic chaos. To handle these issues, in the 1940s, Saarinen put forward the "Theory of Organic Decentralization", which advocates the unitary decomposition of large and crowded cities, and the main strategies are "concentration of the individual's living and working conditions, and decentralization of the city's various activity groups" [7]. The emergence of modern residential units was a direct response to the organic decentralization advocation, since these units successfully decomposed some centralized functions of the city center. Among them, Clarence Perry's "Neighborhood Unit [8], Tripp's "Precincts" [9], and Soviet Union's planning practices of "Microdistrict" [10] had constructed the foundations for the modern residential planning theories.

With the rapid development of the residential planning theories and practices, community parks within residential areas were gaining the increasing concerns in green space planning. Compared to the large urban parks that serve in centralized mode, community parks are smaller and serve in decentralized mode. For example, Perry proposed that community parks should be installed integrally with other public service facilities in "Neighborhood Unit" to fulfill the daily recreational demands of residents [8] (Fig. 1.4). Almost in the same period, the calculation model of "accessibility", which arose from the classical location theory of Greek, was established and widely used in guiding the spatial distribution of public service facilities [11], contributing to a more balanced layout of green spaces over the city.

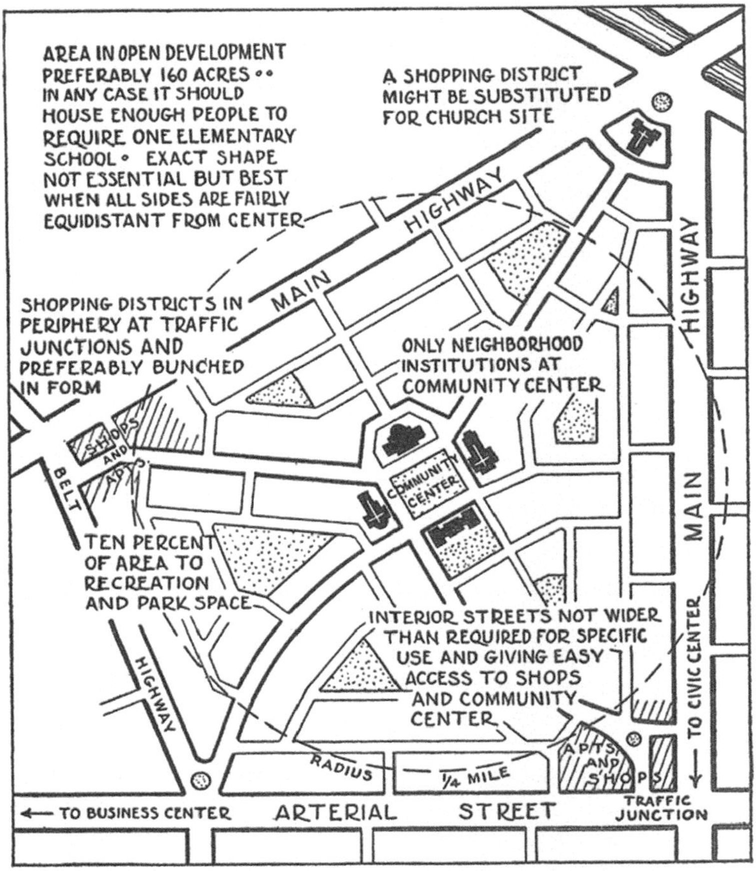

Fig. 1.4 Conceptual model of "Neighborhood Unit". *Source* Perry (1929) [8]

1.1.3 Extensive Renewal

By the 1980s, rapid urbanization had brought about significant distributional disparity of social groups in European and US cities, which exposed vulnerable groups, such as colored minority and low-income cohorts, to the environmental hazards and health risks, leading to severe social contradictions [12]. In the same period, the idea of "Environmental Justice" emerged and rapidly spread as the US civil rights movement rose. The uneven benefit obtained from the green space services of different social groups was regarded as a typical issue of environmental justice [13]. Correspondingly, the large-scale urban renewal practices in European and North American countries were carried out to address the serious inequity in life quality and accessibility of the public services among different groups. Urban green space planning began to shift its focus from system construction to system optimization [14].

To mitigate the inequity led by green space service allocation, new urban green spaces were provided through urban renewal to increase the accessible opportunities for vulnerable groups and enhance the quality of their accessible green spaces, such as the renovation and extension of Riverside Park on Manhattan Island in the 1980s [15]. Meanwhile, residents' travel conditions were ameliorated to expand the accessible scope of the vulnerable groups and enhance their capability to obtain green space services, such as the extensive emergence of greenway and trail networks in European and North American cities at this period [16, 17] [Fig. 1.5].

In this phase, social equity indicators were initially incorporated into urban green space planning to measure the allocational equilibrium of green space service among different social groups. Compared to the baseline quota of land use, although the social equity indicators also measured the quantity of green spaces, the measurement object changed from "land (spatial unit)" to "people (social group)". This broke through the traditional measurement mode that got used to setting the measurement focus on the supply side. Instead, the detailed attributes of demand side were assimilated into the balance process in the decision-making of planning.

1.1.4 Micro-Renewal

In the twenty-first century, as cities' hostile renewal of the developed countries was accomplished or decelerated, their cities entered into the phase focusing on micro-renewal and refined governance. Meanwhile, the "New Public Service" theory, which arose from the theory of democratic citizenship as well as the community and citizenship model, emerged as the dominant guiding ideology in city's public management. Its core idea was to change the government's role from a "Helmsman" to a "Servant", whose main task is set to fulfill the diverse and individualized demands of various social groups [18]. Correspondingly, the life circle system gradually replaced the traditional residential quarters and served as the basic unit in residential planning.

At this background, more concerns and deliberations were given to balance the supply–demand relationship of green space services, aiming to respond to the various demands of different groups and achieve the goal of "public good" [19, 20]. In contrast to the stable and objective criterion of "social equality", "public good" pursues the relative fairness within the specific social conditions and relationships, which is contingent upon the value judgments grounded on the specific demands of each entity and characterized as highly dynamic and subjective. Accordingly, it is commonly measured through the social survey of citizen satisfaction [21]. As the demands for green space services among different groups vary significantly, more precise and fine-grained indicators were applied in green space planning practices to present the satisfaction degree of the service demands of various groups. Compared to the social equity indicators, although the indicators on satisfaction degree also focus on measuring the conditions of demand side, the measurement is built on the premise that different groups hold their unique and individualized demands [18].

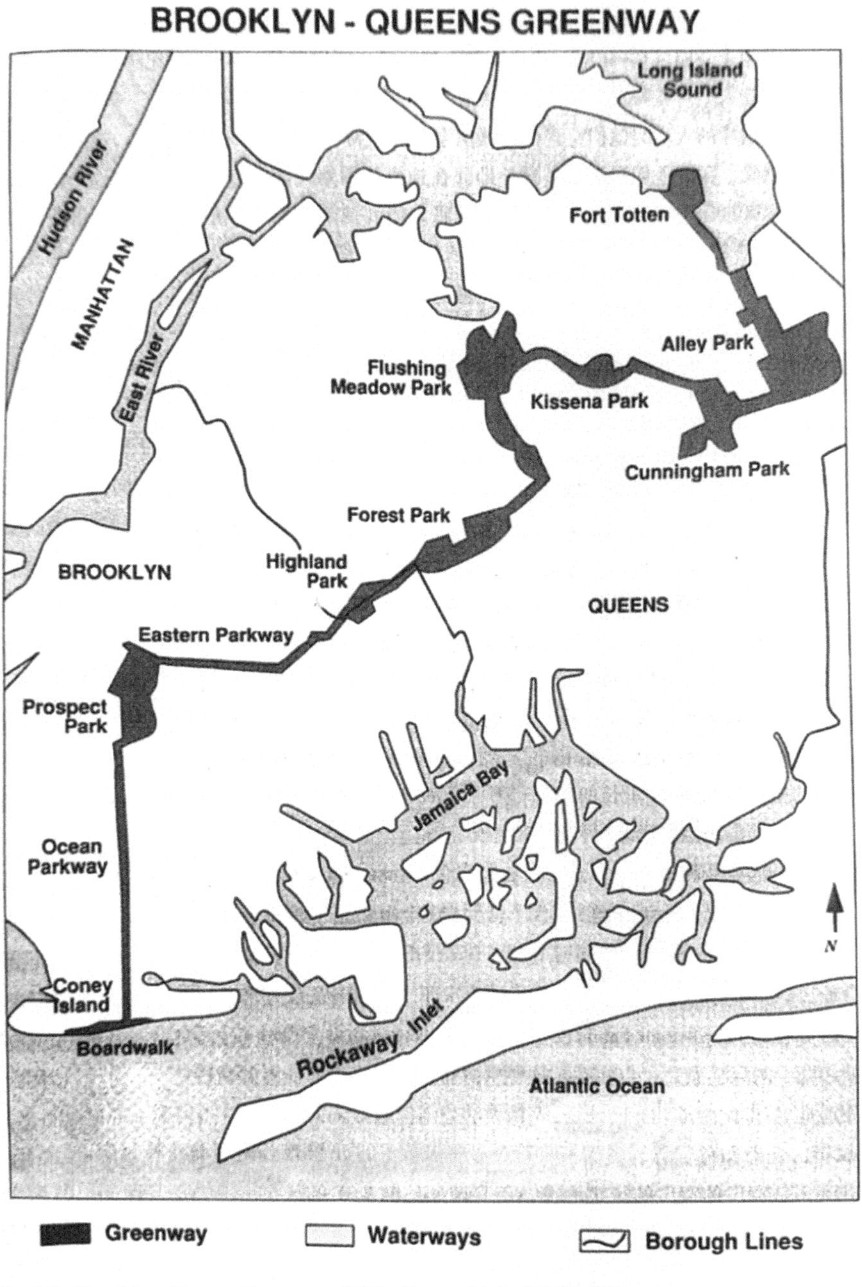

Fig. 1.5 Brooklyn-Queens Greenway, USA. *Source* Little (1995) [17]

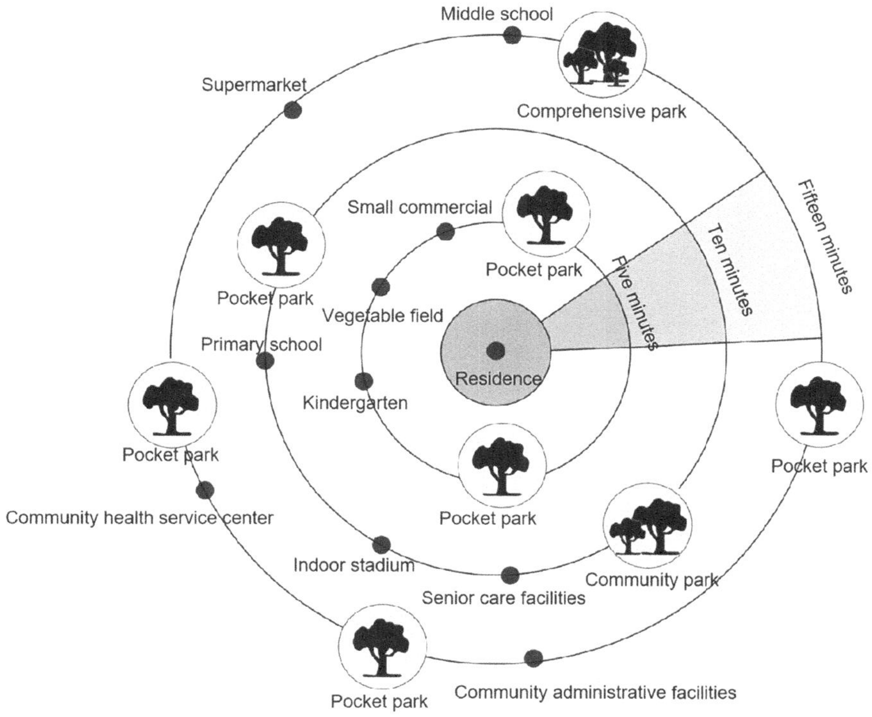

Fig. 1.6 Conceptual model of 15-min life circles

Simultaneously, because the installation of large-sized green spaces is severely limited in built environment [22, 23] [Fig. 1.6], European and US cities began to explore new approaches on supplementing, improving, or replacing urban park services in high-density environment. These initiatives directly led to the popularization of micro green spaces, most of whom are pocket parks (Figs. 1.7, and 1.8). Concurrently, the implementation of time-sharing and other innovative management strategies has also facilitated the development of the micro green spaces within built environment, such as those along or within roads, under highways, and associated with other shared infrastructure [24] [Figs. 1.9, 1.10, 1.11 and 1.12].

1.1.5 Origin of Pocket Parks

Pocket parks, also known as vest pocket parks, is a concept based on the size standard. Their size is usually smaller than the threshold area of the community parks. Similar concepts include micro parks, mini-gardens, tiny parks, etc. [25]. The earliest prototype of pocket parks was conceived in European cities after World War II:

Fig. 1.7 Plan and image of Fortune Street Park (London, UK). *Image source* https://www.spacehive.com/fortunestreetparkfunday

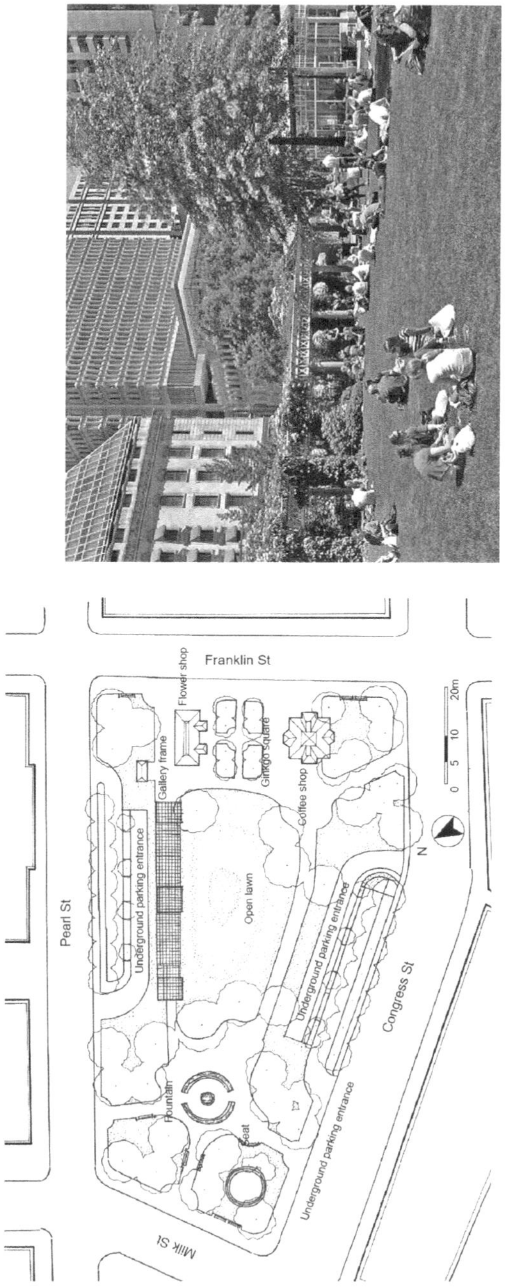

Fig. 1.8 Plan and image of Norman B. Leventhal Park (Boston, USA). *Image source* https://www.tighebond.com/project/norman-b-leventhal-park-at-post-office-square/

Fig. 1.9 Plan and aerial view of Schenley Plaza before renovation (Pittsburgh, USA). *Aerial view source* https://www.sasaki.com/voices/a-decade-of-schenley-plaza-ten-design-takeaways/

Fig. 1.10 Plan and aerial view of Schenley Plaza after renovation (Pittsburgh, USA). *Aerial view source* https://www.sasaki.com/voices/a-decade-of-schenley-plaza-ten-design-takeaways/

Fig. 1.11 Plan and aerial view of Cool Spring Reservoir before renovation (Hermiston, USA). *Aerial view source* Harnik (2010) [24]

Fig. 1.12 Plan and aerial view of Cool Spring Reservoir after renovation (Hermiston, USA). *Aerial view source* https://www.hfadesign.com/cool-spring-reservoir

as cities were reconstructed, some bomb-damaged sites were developed into small parks, such as Ropers Garden in London [26, 27] [Fig. 1.13]. Although the emergence of this unconventional park form was a result of compromise to the serious shortage of capital, labor, and materials at that period, it met with greater success than had been anticipated, contributing significantly to the restoration of cities' physical environment and social spirit [25, 26].

Although the conceptual prototype originated in Europe, the nomenclature and popularization of the "Pocket Park" concept stemmed from the practices in US. In the 1950s, professor Carl Linn from the University of Pennsylvania brought this concept to the United States and cooperated with local governments to create "neighborhood commons" by utilizing the reclaimed tax-delinquent land in Philadelphia, Washington D.C., and Baltimore [26]. In the 1960s, promoted by Mayor John Lindsay's administration and Whitney Seymour, the head of the Park Association of New York, the concept of tiny open spaces and green spaces, whose area could be as small as a single building unit lot (100 ft. × 20 ft., approximately 185 square meters), were strongly advocated in city planning [28]. This approach effectively extended the service of urban parks and relevant recreational resources to those inner-city locations where the density is extremely high and open spaces were severely scarce. It also revolutionized people's traditional recognition of the concept of parks which was rigidly supposed to be large-sized and serve in a centralized mode. Filling interstices between the high-density buildings and adapting to the grid fabric of New York City, such open spaces are characterized as tiny-sized and rectangle-shaped. They are often open to street on one side and enclosed by buildings on the other three sides, like the "pockets" on a jacket [29] [Fig. 1.14]. This is the origin of the name of pocket parks.

The *New York City's Zoning Resolution* enacted in 1961 also took the lead in incentivizing developers to reserve ground sites for public spaces such as plazas and green spaces through the measures like floor area compensation, thus promoting the rapid development of pocket parks at both the institutional and economic levels [30]. In 1963, landscape architect Robert Zion demonstrated his prototypical design of the pocket park system at the exhibition 'New York Parks for New York', organized by the Park Association of New York and staged at the Architectural League [31]. His core proposal was to make up the severe deficiency of outdoor recreational spaces through a systematic transformation of the small derelict plots and interstitial vacant spaces between buildings [32]. By 1965, New York City had built up 18 pocket parks and planned to establish a daily recreational network composed by up to 200 small neighborhood open spaces [26]. In 1967, Paley Park designed by Zion in central Manhattan was built up, which achieved great success and made the concept of pocket park popular over the world [33].

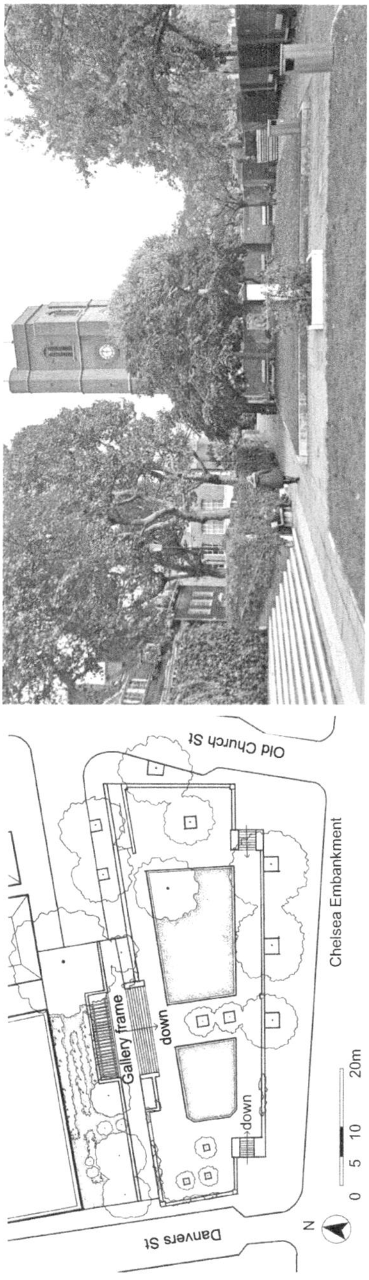

Fig. 1.13 Plan and image of Roper's Garden (London, UK). *Image source* https://www.ianvisits.co.uk/articles/london-pocket-parks-ropers-gardens-sw3-34099/

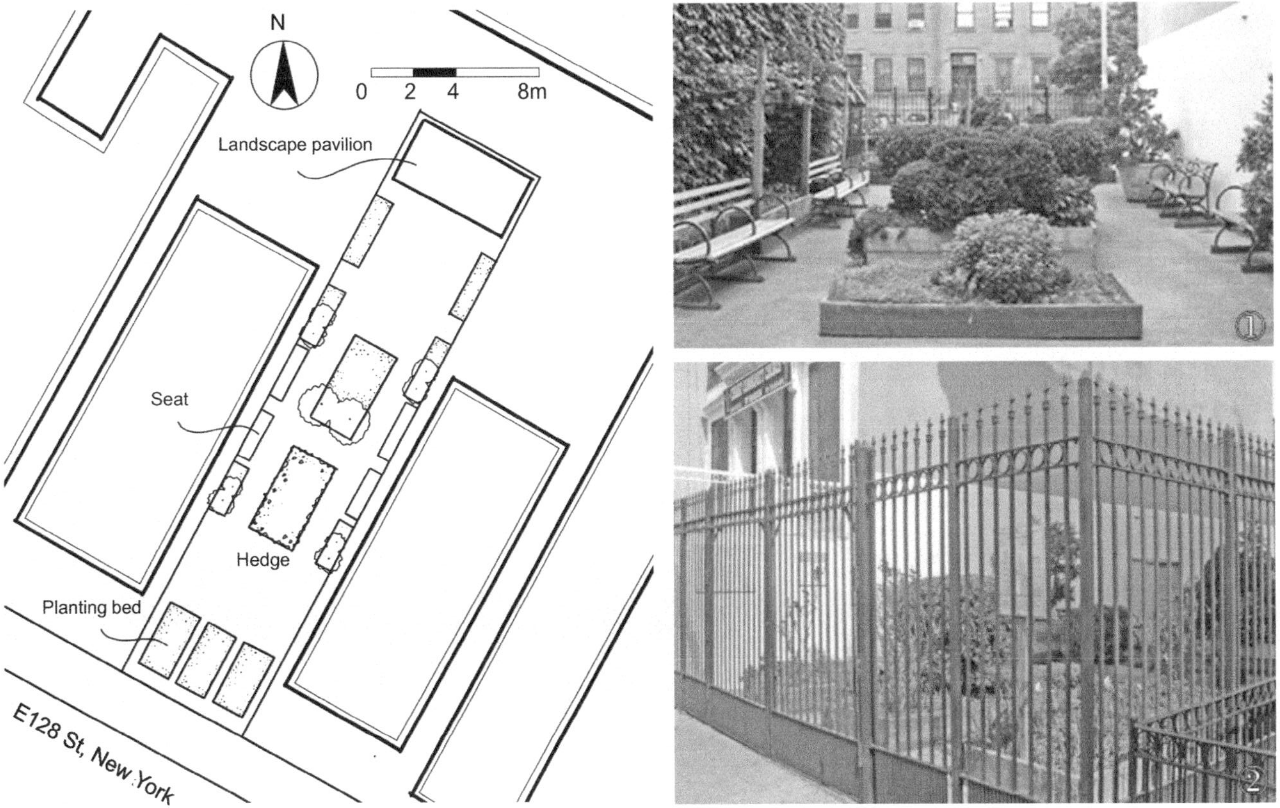

Fig. 1.14 Plan and image of Rev. Linnette C. Williamson Memorial Park (New York, USA). *Image source* ① http://www.williamsonparks.org/vestpocket-park-1965. *Image source* ② Armato (2017) [29]

1.1.6 Development of Micro Green Spaces in China

In the beginning of the 1990s, China's *Code for the Design of Public Park (CJJ48-1992)* listed the park categories of "residential mini-parks" and "streetside mini-parks", which can be classified into micro green spaces in the early phase of China's green space planning [34]. However, in the phase of China's rapid urbanization (1980–2010), the emphasis of green space development was mainly placed on the guarantee of the baseline area and ratio of land resource reserved for green spaces. At this period, micro green spaces, whose contribution to the growth of green space area was minuscule, had not been sufficiently concerned. However, from the beginning of the 2010s, most Chinese cities gradually shifted the development mode from fast expansion to organic renewal and paid more attention to the development quality rather than development speed. The guiding ideologies such as "connotative development" [35], "refined governance" [35], and "dormant land activation" [36] began to dominate the planning and construction practices, for which urban built-up areas became the highlight districts for planning regulation and control. Since micro green spaces showed significant advantages in adapting to the built environment compared to the larger parks (e.g. urban parks and community parks), they can contribute much to filling the gap between the supply and demand of outdoor recreational spaces in the densely populated areas of Chinese cities. This also explains why the planning and development of micro green spaces have gained much more attention at this period.

In 2011, Qiu Baoxing, the vice minister of the Ministry of Housing and Urban–Rural Development, officially introduced the concept of "Micro Green Space" for the first time in China, and proposed that it should play an essential role in the reform of urban development mode led by the urban micro-circulation [37]. In 2017, the Ministry of Housing and Urban–Rural Development issued *Guidelines on promoting the work of Urban Renovation and Ecological Restoration*, advocating to increase new green spaces through demolishing, impermeable ground conversion, and greenery insertion into any possible interstices, creating favorable policy and legislative atmosphere for the development of micro green spaces (Table 1.2) [38]. In 2022, the Ministry of Housing and Urban–Rural Development (MOHURD) of China officially introduced the concept of "pocket park" at the national level through the enaction of the *Announcements on Promoting the Construction of Pocket Parks*, and mandated the provincial and city governments to formulate and implement the development plan of pocket parks [39]. Afterwards, pocket parks gradually became the main tool in supplementing the outdoor recreational opportunities within built-up areas of Chinese cities.

Table 1.2 Concepts and indicators related to pocket parks in China's planning standards and government documents

Enactment time	Standard name	Related concept	Connotation	Indicator requirements
1992	Code for the Design of Public Park (CJJ48-1992)	Residential mini-parks	Parks that should set children's play facilities and respond to the recreational demands of the seniors	Area >0.5 hm^2
		Streetside mini-parks	Parks that should grow exquisite garden plants, emphasizing the artistic effect of the street scenery and providing facilities for short-term rest	Proportion of green spaces ≥ 65%
1993	Code of Urban Residential Areas Planning & Design (GB 50,180–93)	Mini-parks	Parks that serve community residents with public openness	Area 0.4 ~ 1 hm2
2002	Standard for Classification of Urban Green Space (CJJ/T 85–2002)	Residential mini-parks	Green spaces that provide centralized service for the residents of a community	Service radius 0.3 ~ 0.5 km
		Streetside mini-parks	Independent green spaces that located outside the urban road lands, including streetside square, streetside green spaces, etc	Proportion of green spaces ≥ 65%
2016	Code for the Design of Public Park (GB 51,192–2016)	Residential community gardens	Green spaces where children's play facilities must be provided, while the recreation needs of the elderly should be accommodated	Area >0.5 hm^2

(continued)

Table 1.2 (continued)

Enactment time	Standard name	Related concept	Connotation	Indicator requirements
		Street side green space	Green spaces dominated by beautifully arranged garden plants, with artistic streetscape effects and facilities for short breaks should be provided	Proportion of green spaces >65%
2017	Standard for Classification of Urban Green Space (CJJ/T 85–2017)	Mini-parks	Green spaces with independent land parcels, small size, and diverse shapes that can be easily accessed by residents and serve for recreational function	Proportion of green spaces ≥ 65%
2018	Standard for Urban Residential Area Planning and Design (GB 50,180–2018)	5-min pedestrian-scale neighborhood parks	Parklands that are located within the 5-min pedestrian-scale neighborhood and serve for the recreational and sports function for the residents	Area ≥ 0.4 hm2;width ≥ 30 m;internal sports land accounts for 10 ~ 15%
2019	Standard for Planning of Urban Green Space (GB/T 51,346–2019)	Mini-parks	The same as Standard for Classification of Urban Green Space (CJJ/T 85–2017)	Per capita indicator ≥ 1 m2;area 0.1 ~ 1 hm2;service radius 300 m
2022	Announcements on Promoting the Construction of Pocket Parks of the General Office of the Ministry of Housing and Urban–Rural Development (Construction Office 〔2022〕 No.276)	Pocket parks	Greening and activity sites that are open to the public, small-sized, diversified in shape, and used for outdoor recreation	Area 0.04 ~ 1 hm2

1.2 Conceptual Connotations of Pocket Parks

1.2.1 Pocket Parks in Narrow Sense

The original concept of pocket park was primarily used to describe the specific open space generated in the interstices or the courtyards of the high-density buildings of the US cities [40]. It actually integrally summarizes three features of such open spaces in terms of size, shape, and border form, respectively: (1) The size of pocket parks is much smaller than that of traditional urban parks and community parks, like a tiny and exquisite "pocket"; (2) The early pocket parks are commonly developed on the rectangle plots, shaping as a "pocket" of a jacket; (3) Pocket parks are typically embedded within urban blocks, enclosed by buildings on their three sides and open to the street on one side. These border attributes resemble those of the jacket "pocket" that is only open on one side (Fig. 1.15), for which pocket parks are also named as "inner block parks" by Baltimore government in the 1960s [41]. In accordance with this definition, pocket parks are supposed to possess the aforementioned characteristics of size, shape, and border form. Due to the multifaceted features defined in the concept, there are clear restrictions for applying this concept in practice. However, the multiple restrictions also significantly compressed the concept's applicable scope, for which it can be regarded as the narrow sense of the concept of pocket parks.

1.2.2 Pocket Parks in Broad Sense

With the popularization of the utilization of pocket parks in the global planning practices, its conceptual connotation had been gradually expanded and generalized. The most distinct variation throughout this process was that the size feature of the "pocket" remained emphasized, whereas the features of the shape and border form of the "pocket" had been gradually weakened. For example, the *Pocket Park Toolkit*, promulgated by The Trust for Public Land (TPL) of US in 2020, announced that "*The defining characteristic of a pocket park is its size. Although there is no strict definition, Pocket Parks are generally recognized as public park spaces that occupy less than one acre of land (approximately 0.4 hectares).*" [25]

Owing to the high popularity and vividness of the concept of "pocket", pocket parks are often categorized as a type in park classification system of the planning standards of many countries, especially those using size as the main classification criteria. For example, the 2004 edition of *The London Plan: Spatial Development Strategy for Greater London* established a six-class park service system encompassing regional parks, metropolitan parks, district parks, local parks and open spaces, small open spaces, and pocket parks. The area and service radius of the parks decline along with the descending classes. Of these classes, the area of pocket parks is stipulated to be under 0.4 hectares while their service radius is less than 400 m [42].

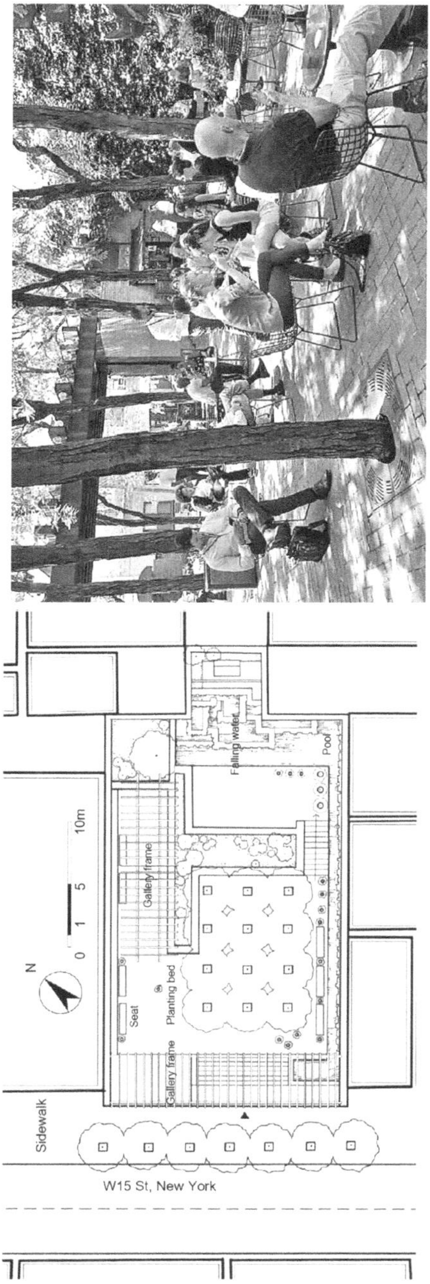

Fig. 1.15 Plan and image of Greenacre Park (New York, USA). *Image source* https://www.tclf.org/sites/default/files/microsites/landslide2017/greenacre-park.html

Upon a meticulous examination of this defining mode of pocket parks, it can be found that, aside from the public nature inherent in parks, the fences that restrict the shape and border form of pocket parks in the early definition had been broken through, while size feature became the only necessary condition in these definitions. It means that public green spaces of various forms, locations, and border forms can be classified as pocket parks in condition that they can meet the area threshold. With fewer and looser constraints, this definition can be regarded as the pocket parks in broad sense.

In this book, we used the broad definition for the concept of pocket parks, as it is more consistent with the existing conditions and development tendency of the planning practices. Since size is the only characteristic emphasized in this definition, the pocket parks in broad sense are totally equivalent to tiny and public green spaces. For example, in the *Announcements on Promoting the Construction of Pocket Parks* issued by the MOHURD of China, pocket parks are defined as "*the greening and activity sites for recreation, which are open to public, small-sized, and multi-shaped, with an area commonly in the range from 400 to 10,000 square meters.*" [39]

1.3 The Patterns of Pocket Parks

According to the origin and conceptual connation of pocket parks, the main task of the pocket park installation is not to provide decorative urban spaces through complex and costly designs, as the space appearance is not the primary concern [29]. La Pietra & Brayer (2011) pointed out that these spaces "*shouldn't 'intimidate' with their decorum and with their refinement of materials which were used, but they should only be welcoming places where everyone could live the experience of being and living at home*" [43]. Therefore, contrary to the traditional concept of "parkland" which simultaneously emphasizes usage, land ownership, and management entity, the concept of "pocket park" attaches the most importance to the actual usage of the space, demonstrating greater inclusiveness and practical effectiveness. Due to the greater flexibility in land ownership and management, the types of green spaces that can be included under the concept of pocket parks are much more extensive than those under the concept of parklands.

However, although pocket parks emphasize the size attribute of green spaces, size is not the sole criterion for determining a pocket park. Referring to relevant planning standards and guidelines over the world, apart from being enough small, pocket parks are also characterized as high level in openness and shareability. This also explains why access-limited green spaces, like private courtyards and internal greenery of fenced neighborhoods, are not listed as pocket parks. By integrating the aforementioned characteristics with the land use flexibility, the full element spectrum of pocket parks can be figured out, not only encompassing the "formal green spaces" which are developed based on the independent parcels of parklands, but also including the "informal green spaces" which are developed on the parcels of other land use categories other than parklands.

1.3.1 Pocket Parks in Formal Green Spaces

Designated with specific public lands for development, formal green spaces are usually invested, constructed, and maintained by the public authorities. According to China's *Code for Classification of Urban Land Use and Planning Standards of Development Land (GB50137-2011)* [44] and *Standard for Classification of Urban Green Space (CJJ/T85-2017)* [45], the scope of formal green spaces covers the traditional categories of "G1 Public Park" and "G2 Green Buffer" as well as the newly increased category "G3 Square" whose main function and greening level are homogenized with conventional green spaces. Any of these formal green spaces that fall into the size range of pocket parks and are freely accessed can be regarded as pocket parks, including small-sized parklands, open-accessed green buffers, and tiny squares.

1.3.2 Pocket Parks in Informal Green Spaces

So far, there is no unified definition for informal green spaces. The "informality" of green spaces are usually explained on three aspects: (1) "non-green space" in predefined land use, indicating that these spaces are not defined as formal green space in land use planning; (2) "greenization" in function, implying that these spaces serve as the green space in residents' daily life; (3) "looseness" in management, denoting that land owners would not overly intervene the daily use of these spaces, generating less restrictions for people's activities than those of the formal green spaces [46–48].

Apart from these three common features under consensus, there remain disparities among different scholars in the concept definition. The disparities often lie in the degree of looseness, publicness, and geographical scope. For example, Rupprecht et al. [48] and Rupprecht and Byrne [46] considered informal green spaces as a type of quasi-public green spaces and restricted their geographical scope to the built environment that has a history of strong anthropogenic disturbance [46, 48]. Moreover, the appearance of non-remnant and spontaneous vegetation is also set as one of the core conditions in defining informal green spaces. As this definition mode sets very stringent conditions for identifying informal green spaces, it can be regarded as the informal green spaces in narrow sense.

Contrarily, Sikorska et al. [51] and Włodarczyk-Marciniak et al. [50] contended that informal green spaces include public green spaces and non-public green spaces such as private gardens and other forms of green spaces within access-limited areas [50, 51]. The geographical scope is also expanded to the natural environment for which farmlands and reserved areas are included in the concept. Furthermore, in their opinion, the "looseness" of management and maintenance merely exists in the level of usage and management, that is, non-remnant and spontaneous vegetation is not seemed as a necessary attribute for informal green spaces. Since this definition is relatively lenient in determining the conditions and nature of informal green space, it can be regarded as the informal green space in broad sense.

In China's planning system, all the informal green spaces fall into the category of "XG Attached Green Space" in the classification system of urban green spaces, which do not own an independent land parcel. By jointly comparing the existing definition modes of informal green spaces, the proper definition that is suitable for China's planning system should focus on the elements available in planning control and regulation, that is, the attached green spaces that own high publicness or potential to be open to public in the built environment. This description is more closed to the narrow mode of Rupprecht's definition, which emphasizes geographical scope of the built environment and the publicness of spaces. However, as the development intensity and saturation are high in China's built-up areas, the proportion of abandoned lands, vacant lands, and other unmanaged lands is relatively low. If the feature of "non-remnant and spontaneous vegetation" is insisted in the informal green space definition in China, the scope of regulatable green spaces will be significantly narrowed, greatly weakening the value and effectiveness of the informal green space planning. To maximumly integrate the potential resources for optimizing the green space service in planning, the requirements for the degree of "looseness" in management of Sikorska's definition can be combined with the requirements of other aspects of Rupprecht's definition (Tables 1.3, 1.4).

The informal green spaces, which have an area of 400–10,000 m^2 and can be used for public recreation, are also seemed as pocket parks. In practice, we can find that the majority of pocket parks fall into the category of informal green spaces, including green spaces attached to the lands of residential, administration and public services, commercial and business facilities, road, street and transportation, and so on. Since the total area of attached green spaces accounts for a considerable proportion of the total green space area of a city (usually more than 50%), they are also becoming the main source of potential lands for pocket park development.

1.4 The Value of Pocket Parks

1.4.1 City Level

At the city level, the extensive installation of pocket parks can significantly increase recreational opportunities in severely limited conditions of high-density areas and extend the green service network to the areas that the service of conventional parks cannot reach, which makes them an effective tool for green service relief and equitable allocation of public resources. Some people also vividly titled pocket parks as "Green Stepping Stones" in the urban fabric, that is, *"they could function as a compliment to larger parks and to some extent fill the need for peoples every day contact with nature"* [52] [Fig. 1.16].

Table 1.3 The concept definition mode and characteristics of informal green spaces

Concept	Elements covered	management and maintenance forms		Locations		Openness	
		Officially-managed	Non-managed	Built environment	Natural environment	Publicity	Non-publicity
Informal green spaces in the narrow sense	Street side spaces, vacant lands, interstitial spaces, spaces along railway, brownfields, waterfront spaces, open spaces around facilities, micro-sites, open spaces along power lines	–	●	●	–	●	–
Informal green spaces in the broad sense	Green spaces in natural reserves, vacant lands, open spaces in industrial and post-industrial areas, grasslands and farmlands, waterfront open spaces, open spaces within social service areas, recreation areas, public service areas, and residential areas	●	●	●	●	●	●

(continued)

Table 1.3 (continued)

Concept	Elements covered	management and maintenance forms		Locations		Openness	
		Officially-managed	Non-managed	Built environment	Natural environment	Publicity	Non-publicity
“Attached green spaces” in China’s “Standard for Classification of Urban Green Space”	Green spaces attached to residential areas, public management and public service facilities, commercial service facilities, industry, logistics and warehousing, road and traffic facilities, and public facilities	●	●	●	–	●	●
Informal green spaces in China’s planning system	Attached green spaces with public openness or potential to open	●	●	●	–	●	–

Table 1.4 Comparisons between formal green spaces and informal green spaces

Attribute	Formal green space	Informal green space
Land use category	Green space and Square	Various land use categories out of the Green space and Square
Green space types (code)	Public Parks (G1), Green Buffers (G2), Squares (G3)	Attached Green Spaces (XG)
Individual size	larger than 0.1 hm^2	Tiny-sized (commonly smaller than 1 hm^2)
Land use form	Independent land use plot	Non-independent land use plot
Spatial definition	Clear boundaries	Ambiguous boundaries and strong liminalities
Custodial department	Park, gardening, and recreation management departments	Entities or individuals that own the right of use or ownership the lands
Management and maintenance degree	Strict	Looser

In China, given the high density of most cities, the scarcity of recreational resources in high density cities leads to an attenuation on the hierarchical characteristics of urban green spaces in daily recreational services and an intensification of the tendency of flattened service (Fig. 1.17). This tendency became increasingly prominent after the abolition of entry fees for comprehensive parks over the country at the head of the 2000s. This approach gradually transformed comprehensive parks, which were originally set as the weekend recreational destinations, into important daily recreational places for nearby residents. In this process, the emergence of pocket parks not only provides an important alternative to compensate for the absence of conventional parks in optimizing the functions and spatial structures of urban green spaces, but also offers a solid guarantee for the well-beings of residents' lives as well as the sustainability of urban development.

Meanwhile, the development of pocket parks can also reactivate abundant negative spaces that are dormant, abandoned, or inefficiently-used, thereby serving as a catalyst in the regeneration of urban areas in a large scale. This operational mechanism is also designated as the interventions of urban microsurgery, namely "*an extremely precise work on very restricted surfaces* [29]". The principle of this work is to adopt pocket park as a planning tool to precisely heal the problematic spots through small-scale and customized projects in the built environment and stimulate the reactivation of the larger urban structures, achieving an effect of "moving thousand cattle with four tales" [53].

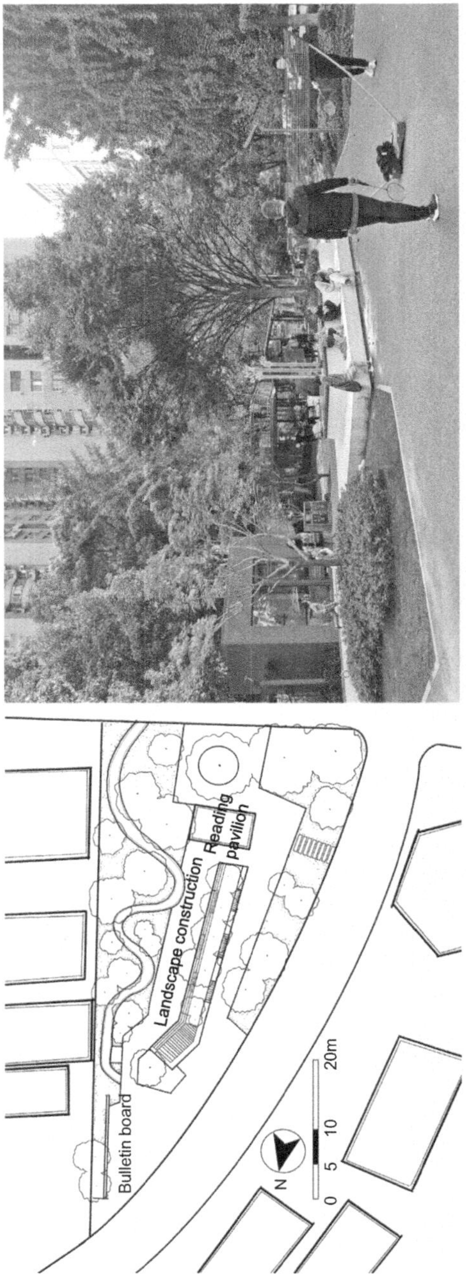

Fig. 1.16 "Green stepping stones": Plan and image of Red Star Square (Nanjing, China)

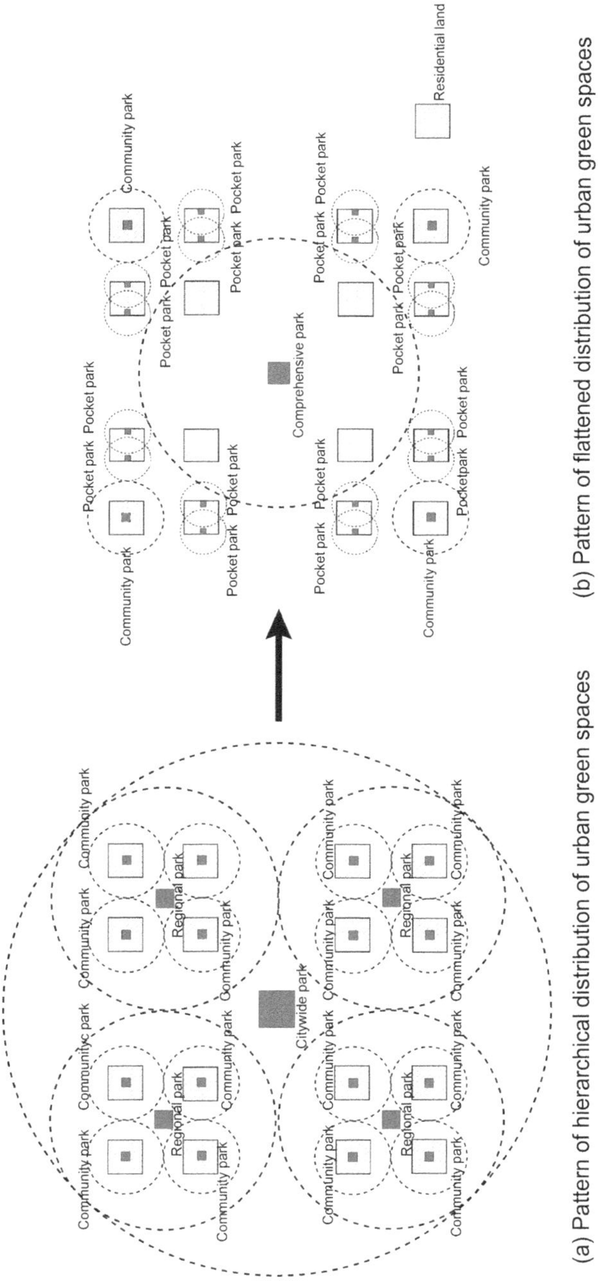

Fig. 1.17 Transformation of urban green space distribution pattern

1.4.2 Community Level

In high-density communities characterized by high building density and seriously lack of outdoor recreational spaces, tiny-sized pocket parks showed better adaptability to the built environment and higher installation feasibility compared to larger conventional parks (comprehensive parks, community parks, etc.). In these communities, pocket parks are often regarded as precious "outdoor living rooms", serving as essential places for community residents to intimately contact with each other and increase social connections. They can effectively enhance the sense of belonging and well-being of residents in the modern cities and mitigate the psychological issues led by social alienation [29, 54, 55] [Fig. 1.18]. Charles Taylor once described the spaces within pocket parks as "*A space where people could exchange the knowledge of the habits and customs of the others, not be afraid of being influenced by other cultures to the point of suffering the loss of their own identity*" [29].

Meanwhile, pocket parks can also create a series of proximate and convenient outdoor recreation opportunities for residents. This precious proximate convenience in high-density environment can save residents' time for outdoor traveling and evoke their willingness for recreational outings. This is very important for maintaining the physical and mental health of middle-aged and young people in the modern society [56]. For the seniors or other people with restricted mobility, proximate pocket parks also reduce the frequency of the trunk road crossings in their daily lives, through which the risk of traffic accident for them can be significantly lower down [57].

1.4.3 Other Values

With the worldwide exacerbation of the ecological and environmental challenges in recent decades, the function of pocket parks in ecological protection and environmental governance have also received increasing concerns. For example, existing studies find that pocket parks can provide foods and habitats for creatures such as birds, pollinating insects, small reptiles, and mammals within cities, thereby enhancing the species diversity within cities [58–60]. The soft ground covers (such as turf and water) and natural shades within pocket parks can effectively reduce the surface temperature of the surrounding areas and improve the local microclimate [54, 61]. If the planning of the pocket park system can be integrated into the urban stormwater control system, they can also play positive roles in the runoff detention and stormwater storage, thereby reducing the risk of urban waterlogging [62, 63].

Fig. 1.18 Plan and image of Shuiyougang Pocket Park (Nanjing, China)

1.5 Issues and Challenges in Pocket Park Planning and Design

1.5.1 Unique Attributes

Compared to the conventional large parks, pocket parks demonstrated particularities in their inherent attributes and service mechanisms. Accordingly, traditional park planning and design guidelines—which were mainly derived from studies on larger parks' service mechanism—may not account for the real conditions and requirements of pocket park planning and design. For example, the pocket parks' tiny size restricts their service capacity, for which they cannot directly copy the centralized service mode of the large-sized urban parks and have to serve in a decentralized mode instead (Fig. 1.19). This further results in the low applicability of the conventional "top-down" planning paradigm in pocket park planning. Due to the lack of the systematic and in-depth knowledge of their unique attributes and service mechanisms, it is difficult to guarantee the effectiveness of current planning and design practices of pocket parks, frequently leading to the issues like ambiguity in functional positioning, randomness in site selection, and weakness in parks' collaboration. This might leave behind a series of hidden risks, such as unbalanced distribution, service dysfunctional, and land resource waste.

1.5.2 Adaptation in Built Environments

The planning and design practices of pocket parks are mostly carried out in high-density environment. Various existing elements of built environment, such as space, society, and culture, might influence the planning and design of pocket parks. These influences mainly act on two aspects: parks' site selection and service performance. As the space available for pocket park installation is strictly limited in built environment, factors such as the existing urban texture, land use structures, and land ownerships directly affect pocket parks' site selection and installation feasibility. Meanwhile, due to the limited size, some internal amenities and facilities that can be easily installed in larger parks, such as pathways, water features, cafes, retail outlets, and toilets, are mostly absent in pocket parks. Therefore, compared to larger parks, the integrity and density of surrounding recreation-related facilities play more important roles in supporting the pocket park service, generating a result that pocket park's service performance is much more sensitive to surroundings. If planners and designers aim to reach a high level of parks' service performance through a cost effective approach, it is necessary to fully "leverage" the existing environment in pocket park planning and design. This requires for, on one hand, a sufficient integration with the positive factors in the built environment to facilitate the installation and service of pocket parks; and on the other hand, shielding against the interference

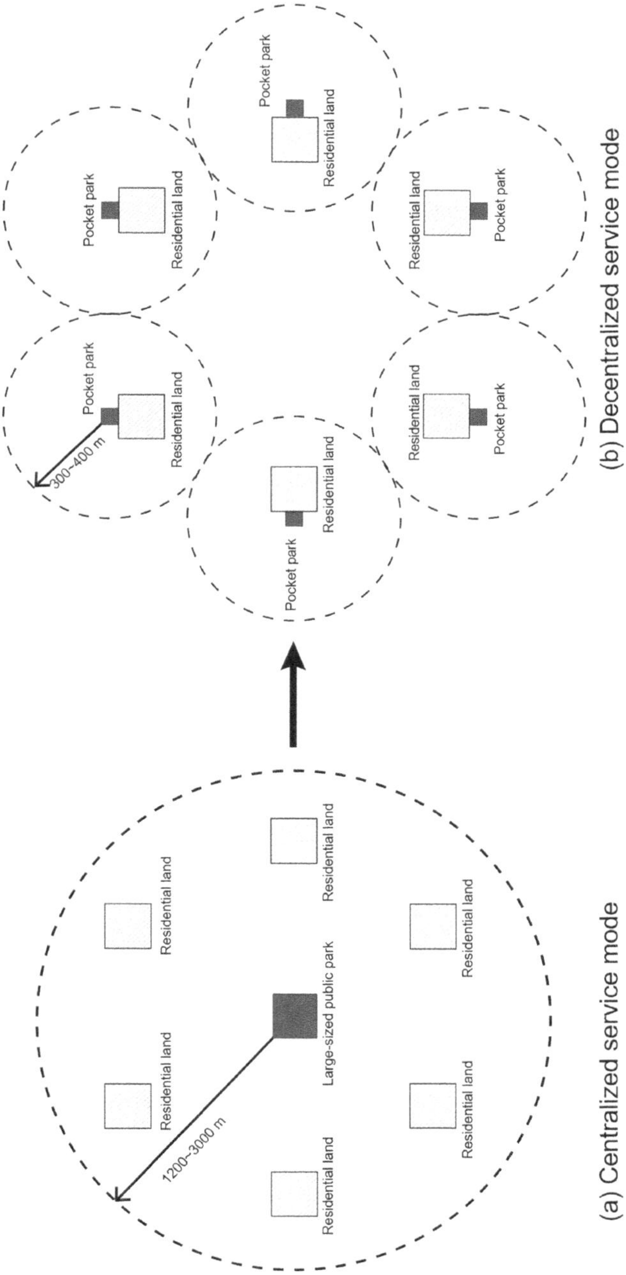

Fig. 1.19 Transformation of urban green space service pattern

of the negative factors. However, as the influencing factors are numerous while their action ways are diverse, it is still a great challenge to make a targeted control of these factors in the planning practice.

1.5.3 Collaboration of Multiple Parks

Due to the limited capacity of a single pocket park, it is difficult to effectively make up the large gap between the supply and demand of outdoor recreational resource in high-density areas through the individualized and randomized mode of pocket park development. Therefore, in the early stage of the pocket park development in US cities around the 1960s, a strategy of establishing a pocket park system to compensate for the absence of conventional parks in high-density communities was proposed [26]. Nowadays, facing to the sustained rise of the urbanization rate and population density, establishing "collaborative" pocket park groups to efficiently respond and decentralize the intense and centralized recreational demands becomes an attractive and effective solution for decision-makers. However, in the service process, apart from collaborating with each other complementarily, proximate pocket parks may possibly compete with each other which will inhibit parks' service performance [24, 27]. Relevant studies also showed that high homogeneity in the functional positioning, design style, and facility configuration of proximate green spaces may intensify this competitive effect and compromise the efficiency of resource allocation [59]. Therefore, in the planning and design process, it is not enough to just find a proper site for each pocket park to install at the individual level, but also essential to guide and regulate the interaction modes of multiple pocket parks at the group level, by which the collaboration among them can be activated while their competition can be avoided or minimized.

1.5.4 Control and Regulation Methods

The transition of urban development mode from fast expansion into organic renewal makes small-sized green spaces the main objects in current planning and design practice. Accordingly, the planning focus is being converted from the construction or hostile optimization of green space system into system micro optimization. The conversion of the planning focus also leads to a series of variations in the planning goals, regulatory indicators, methods and techniques, and guarantee approaches. For example, traditional planning paid much attention to the area growth of green spaces to ensure enough quantity of green spaces can be provided to support the fast urbanization. However, as the individual size of pocket park is limited, their contribution to the growth of the total area of green spaces is much less than that to the growth of the recreational opportunities, accessibility, and equity. Therefore, traditional park planning and design indicators—which are mainly used to guide the larger park development—may not be equivalently essential and applicable in pocket park planning and design. Meanwhile, the uniqueness of pocket parks in land use, service mode, and maintenance also requires for the different regulation approaches and implementation measures in planning practices, bringing new challenges for planners and decision-makers who have already been accustomed to formulate planning and design using traditional paradigms.

References

1. Lohrberg F (2001). Peri-urban agriculture in urban and open space planning. https://doi.org/10.18419/opus-17
2. Harigaya S (1991) The history of Western gardening: from the Garden of Eden to natural parks. China Architecture & Building Press. https://doi.org/10.11501/12422974
3. GraftinGardeners (2024, March 29) The history of London's commons. GraftinGardeners. Accessed December 15, 2024, from https://www.graftingardeners.co.uk/history-londons-commons/
4. Howard E, Osborn FJ (1965) Garden cities of to-morrow, 11th edn. Press, M.I.T
5. Corbusier L, Coltman D, Knight P, Levieux E (1964) The radiant city: elements of a doctrine of urbanism to be used as the basis of our machine-age civilization. Orion P. https://books.google.com.hk/books?id=7aqvtAEACAAJ
6. Kowsky FR (1987) Municipal parks and city planning: frederick Law Olmsted's Buffalo park and parkway system. J Soc Archit Hist 46(1):49–64. https://doi.org/10.2307/990145
7. Saarinen E (1943) The city, its growth, its decay, its future. Reinhold Publishing Corporation. Accessed December 10, 2024, from http://archive.org/details/cityitsgrowthits0000saar_o2h1
8. Perry CA (1929) The neighbourhood unit. Routledge/Thoemmes Press. https://books.google.com.hk/books?id=f45enQEACAAJ
9. Tripp HA (1942) Town planning and road traffic. E. Arnold & Company.
10. Konysheva E (2023) Formation of the concept of a residential microdistrict in the Soviet urban planning of the 1930s. Sci J "Academia Architect Construct, 82–92. https://doi.org/10.22337/2077-9038-2023-4-82-92
11. Liu X (2007) General description of spatial accessibility. Urban Transport of China 5(6):36–43
12. Mohai P, Pellow D, Roberts JT (2009) Environmental justice. Annu Rev Environ Resour 34:405–430. https://doi.org/10.1146/annurev-environ-082508-094348
13. Zhang T, Yue Y (2019) A brief review of the English literature on "landscape justice" 1998–2018. Chinese Landscape Architect 35(05):5–12. https://doi.org/10.19775/j.cla.2019.05.0005.
14. Gan W, Wu Y, Zhou Y (2019) The urban renewal strategy of American cities from the perspective of Rawl's theory of justice—An empirical study of Atlanta BeltLine. Chinese Landscape Architecture 35(10):77–82
15. Grimm E, Schroeder EP (2007) Riverside park: the splendid sliver. Columbia University Press.
16. Fabos JG (1995) Introduction and overview: The greenway movement, uses and potentials of greenways. Landsc Urban Plan 33(1.3):1–13. https://doi.org/10.1016/0169-2046(95)02035-R
17. Little CE (1995) Greenways for America (Johns Hopkins paperback ed.). Johns Hopkins University Press
18. Denhardt RB, Denhardt JV (2000) The new public service: Serving rather than steering. Public Adm Rev 60(6):549–559. https://doi.org/10.1111/0033-3352.00117
19. Dai D (2011) Racial/ethnic and socioeconomic disparities in urban green space accessibility: Where to intervene? Landsc Urban Plan 102(4):234–244. https://doi.org/10.1016/j.landurbplan.2011.05.002
20. Tappert S, Klöti T, Drilling M (2018) Contested urban green spaces in the compact city: the (re-)negotiation of urban gardening in Swiss cities. Landsc Urban Plan 170:69–78

21. Li Y, Zhu D, Wang H (2016) Choice of new public service-oriented development mode of urban historic districts: an empirical study of residents' satisfaction in Shanghai City. City Plann Rev 2016(02):51–60
22. Coutts C, Miles R (2011) Greenways as green magnets: the relationship between the race of greenway users and race in proximal neighborhoods. J Leis Res 43(3):317–333. https://doi.org/10.1080/00222216.2011.11950239
23. Xiao X, Li M (2015) Research on the increment of micro-green space in high-density city of Macao Peninsula. Urban Plan Forum 2015(05):105–110
24. Harnik P (2010) Urban green: innovative parks for resurgent cities. Island Press. Accessed from https://search.library.wisc.edu/catalog/9910092652002121
25. The Trust for Public Land (2020) Pocket park toolkit. The Trust for Public Land. Accessed April 15, 2024, from https://www.tpl.org/resource/pocket-park-toolkit
26. Faraci P (1967) Vest pocket parks. American Society of Planning Officials
27. Ibes DC (2015) A multi-dimensional classification and equity analysis of an urban park system: a novel methodology and case study application. Landsc Urban Plan 137:122–137. https://doi.org/10.1016/j.landurbplan.2014.12.014
28. Seymour WN (Ed) (1969) Small urban spaces: the philosophy, design, sociology, and politics of vest-pocket parks and other small urban open spaces. New York University Press. Accessed from http://lib.ugent.be/catalog/rug01:000964963.
29. Armato F (2017) Pocket park: Product urban design. Des J 20(sup1):S1869–S1878. https://doi.org/10.1080/14606925.2017.1352705
30. Li S, Chen L (2018) 1961 New York City zoning resolution. South China University of Technology Press
31. Bernstein FA (2017) Side pocket: New York's Paley Park, which turns 50 this month, is a masterpiece—proof that even a tiny public space can make a difference in a crowded city. Landscape Architecture 107(5):122–129
32. Tate A (2004) Great city parks. Spon Press. Accessed from https://search.library.wisc.edu/catalog/9910478597202121.
33. Zion R (1969) Parks where the people are–the small midtown park. New York University Press.
34. Ministry of Housing and Urban-Rural Development, PRC (1992) Code for the design of public park: CJJ48–1992
35. People's Daily (2015) The Central Urban Work Conference was held in Beijing, and Xi Jinping and Li Keqiang delivered important speeches. People's Daily. Accessed December 10, 2024, from https://www.gov.cn/xinwen/2015-12/22/content_5026592.htm
36. Xinhua News Agency (2024, July 21) Decision of the Central Committee of the Communist Party of China on further deepening reform and promoting Chinese path to modernization. Xinhua News Agency. Accessed December 10, 2024, from https://www.gov.cn/zhengce/202407/content_6963770.htm
37. Qiu B (2011) Rebuild urban micro-cycle—An emerging grand trend. Urban Development Studies 18(08):1–13
38. Ministry of Construction of the People's Republic of China (2017) Guidelines on promoting the work of urban renovation and ecological restoration of the Ministry of Housing and Urban-Rural Development
39. Ministry of Housing and Urban-Rural Development, PRC (2022) Announcements on promoting the construction of pocket parks: [2022] No. 276. Ministry of Housing and Urban-Rural Development, PRC. Accessed from https://www.gov.cn/zhengce/zhengceku/2022-08/09/content_5704766.htm
40. London Parks & Gardens (2024, December 10) Roper's Gardens. London Parks & Gardens Trust. Accessed from https://londongardenstrust.org/conservation/inventory/site-record/
41. Baltimore Heritage. (2014, March 17) Harlem Park. Baltimore Heritage. Accessed December 10, 2024, from https://baltimoreheritage.org/programs/harlem-park/.
42. Greater London Authority (2004) The London plan: spatial development strategy for Greater London. Greater London Authority

43. La Pietra U, Brayer MA (2011) Abitare la città: Ricerche, interventi, progetti nello spazio urbano dal 1960 al 2000. Allemandi
44. Ministry of Housing and Urban-Rural Development, PRC (2011) Code for classification of urban land use and planning standards of development land: GB50137–2011
45. Ministry of Housing and Urban-Rural Development, PRC (2017) Standard for classification of urban green space: CJJ/T85–2017
46. Rupprecht CDD, Byrne JA (2014) Informal urban greenspace: a typology and trilingual systematic review of its role for urban residents and trends in the literature. Urban Forestry & Urban Greening 13(4):597–611. https://doi.org/10.1016/j.ufug.2014.09.002
47. Shaw P, Hudson J (2009) The qualities of informal space: (Re)appropriation within the informal, interstitial spaces of the city. In Proceedings of the conference occupation: negotiations with constructed space (University of Brighton)
48. Rupprecht CDD, Byrne JA, Ueda H et al (2015) 'It's real, not fake like a park': residents' perception and use of informal urban green-space in Brisbane, Australia and Sapporo, Japan. Landsc Urban Plan 143:205–218. https://doi.org/10.1016/j.landurbplan.2015.07.003
49. Sikorska D, Łaszkiewicz E, Krauze K et al (2020) The role of informal green spaces in reducing inequalities in urban green space availability to children and seniors. Environ Sci Policy 108:144–154. https://doi.org/10.1016/j.envsci.2020.03.007
50. Włodarczyk-Marciniak R, Sikorska D, Krauze K (2020) Residents' awareness of the role of informal green spaces in a post-industrial city, with a focus on regulating services and urban adaptation potential. Sustain Cities Soc 59:102236. https://doi.org/10.1016/j.scs.2020.102236
51. Sikorska D, Łaszkiewicz E, Krauze K et al (2020) The role of informal green spaces in reducing inequalities in urban green space availability to children and seniors. Environ Sci Policy 108:144–154. https://doi.org/10.1016/j.envsci.2020.03.007. (Duplicateentry;see#49)
52. Nordh H, Østby K (2013) Pocket parks for people – A study of park design and use. Urban Forestry & Urban Greening 12(1):12–17. https://doi.org/10.1016/j.ufug.2012.11.003
53. Akkerman A, Cornfeld AF (2009) Greening as an urban design metaphor: looking for the city's soul in leftover spaces. Structurist 49:30–35
54. Barton H, Grant M, Guise R (2010) Shaping neighbourhoods: for local health and global sustainability, 3rd ed. Routledge. https://www.taylorfrancis.com/books/9781000403794. https://doi.org/10.4324/9780429321245.
55. Gibson H, Canfield J (2016) Pocket parks as community building blocks: a focus on Stapleton. CO. Community Development 47(5):732–745. https://doi.org/10.1080/15575330.2016.1220965
56. Giles-Corti B, Broomhall MH, Knuiman M et al (2005) Increasing walking: how important is distance to, attractiveness, and size of public open space? Am J Prev Med 28(2):169–176. https://doi.org/10.1016/j.amepre.2004.10.018
57. Sun X, Wang L, Wang F et al (2020) Behaviors of seniors and impact of spatial form in small-scale public spaces in Chinese old city zones. Cities 107:102894. https://doi.org/10.1016/j.cities.2020.102894
58. Strohbach MW, Lerman SB, Warren PS (2013) Are small greening areas enhancing bird diversity? Insights from community-driven greening projects in Boston. Landsc Urban Plan 114:69–79. https://doi.org/10.1016/j.landurbplan.2013.02.007
59. Danford RS, Strohbach MW, Warren PS et al (2018) Active greening or rewilding the city: how does the intention behind small pockets of urban green affect use? Urban Forestry & Urban Greening 29:377–383. https://doi.org/10.1016/j.ufug.2017.11.014
60. Phillips D, Lindquist M (2021) Just weeds? Comparing assessed and perceived biodiversity of urban spontaneous vegetation in informal greenspaces in the context of two American legacy cities. Urban Forestry & Urban Greening 62:127151. https://doi.org/10.1016/j.ufug.2021.127151
61. Bartesaghi-Koc C, Osmond P, Peters A (2019) Spatio-temporal patterns in green infrastructure as driver of land surface temperature variability: the case of Sydney. Int J Appl Earth Obs Geoinf 83:101903. https://doi.org/10.1016/j.jag.2019.101903

62. Jim CY, Chen SS (2003) Comprehensive greenspace planning based on landscape ecology principles in compact Nanjing city. China. Landscape and Urban Planning 65(3):95–116. https://doi.org/10.1016/S0169-2046(02)00244-X
63. Andersson E, Langemeyer J, Borgström S et al (2019) Enabling green and blue infrastructure to improve contributions to human well-being and equity in urban systems. Bioscience 69(7):566–574. https://doi.org/10.1093/biosci/biz058

Chapter 2
Pocket Park Attributes

The basic attributes of pocket parks can be literally decomposed into two aspects: "pocket" and "park". "Pocket" represents the size attribute, that is, it occupies a tiny area, typically smaller than the threshold area of community parks (1 hectare in China). "Park" is the functional attribute, indicating that it should be freely accessed for public recreation as a park. Concurrently, as urban green spaces, pocket parks should at least partially be covered by greenery and be installed on the urban ground, which can differentiate them from the non-greenery activity venues as well as other greenery forms like vertical greening, roof gardens, and balcony greening. The clarification of the basic attributes of pocket parks provides a base for a further discussion on their other extended attributes on functions, spaces, common types, and forms.

2.1 Functional Attributes

2.1.1 Functions

The rapid development of pocket parks stems from the exploration on expanding recreational spaces within the high-density built environment. The report No. 229 *Pocket Parks,* released by the American Society of Planning Officials (the predecessor of the "American Planning Association") in 1967, asserted that "*Although limited both in scope and size, these parks represent a serious effort to improve the quality of the environment in the more crowded urban areas... Their impact, however, is likely to be greatest in those low-income, densely populated neighborhoods where outdoor public space is severely limited.*" [1]. Therefore, till today, the majority of studies and planning practices on pocket parks also set the focus on their function of recreational service in high-density environments.

Centering on the recreational service, pocket parks have also derived other extension functions, such as enhancing communities' identifiability and sense of

C. Zhou, *Principles and Methods of Pocket Park Planning and Design*, Urban Sustainability, https://doi.org/10.1007/978-981-96-5801-5_2

belonging, reducing the crime rates and the incidence of physical and mental illness, and promoting urban innovation (Table 2.1). Apart from these recreation-related functions, research on pocket parks' other functions also increased in recent years, such as biodiversity maintenance [2], local climate regulation [3], and stormwater control [4], which can be uniformly categorized as the non-recreational or non-dominant functions of pocket parks. Accordingly, it is necessary to clarify that the principles and methods of pocket park planning and design discussed in this book mainly focus on the arrangement of the dominant function—recreational service of pocket parks.

It should be also noted that pocket parks are not cure-all for the high-density cities as they cannot completely substitute for the roles of large-sized urban green spaces in the fields of ecological protection, environmental improvement, and some specific recreational services (e.g. camping, boating, and ball games). Pocket parks' size also significantly limits their service capacity to a large extent. Therefore, for either the

Table 2.1 Functions of pocket parks

Categories	Subcategories	Details
Dominant function	Recreational service	Providing proximate recreational sites and facilities for residents of all ages
Derivatives of the dominant function	Enhancing community identifiability	Showing the distinctive features of the communities, there by shaping and highlighting the unique identity of the community
	Enhancing Community's sense of belongingness	Providing the community with a comfortable place for intimate communication of various groups, through which the interactions between residents can be enhanced
	Reducing crime rates	Activating the negative, unused, or inefficient spaces to reduce the chances of offenses, and increasing the sense of security
	Reducing physical and mental illness incidence	Providing residents with convenient fitness venues and facilities to enhance residents' willingness and time to exercise, thereby improving residents' physical health and alleviating residents' anxiety and mental stress
	Promoting urban renewal	Rehabilitating localized urban sites to bring new vitality to the surrounding lots, facilitating the urban renewal in the broader area
Other functions	Maintaining biodiversity	Providing forage and habitat places for insects, birds, and other small animals
	Regulating microclimate	Reducing heat radiation through shrubs, trees, and permeable surfaces and mitigating the heat island effect of the surrounding lots through vegetations and their shades
	Controlling stormwater	Collecting and purifying rainwater by installing bioswales, filtrating equipment, storage facilities, and other measures

entire city or localized areas with concentrated recreational demands, pocket parks are mostly regarded as an effective supplement to conventional parks, which should be integrated into the conventional park system to jointly respond to the residents' recreational demands.

2.1.2 Modes

Unlike large-sized urban parks that adopt a long-radius and centralized service mode, pocket parks mainly adopt a short-radius and decentralized service mode, which can offer proximate outdoor recreational opportunities through an even and dispersed distribution. However, although different countries may set different criterion in defining "proximity", there is a common sense that the proximity should be within the "5-min walking distance (200 ~ 400 m)". For example, in the 1965 edition of *Standards for Outdoor Recreational Areas* released by the American Planning Association, the service radius of playlots, which is the lowest class of the hierarchical system of outdoor recreational areas, is set as 200 ~ 400 m [5]. In the 2004 edition of *The London Plan: Spatial Development Strategy for Greater London,* the service radius of pocket parks and small open spaces is set as no more than 400 m [6]. In Japan, the service radius of block parks, which are equivalent to the category of pocket parks in other countries, was set as 250 m [7]. Similarly, in the 2019 version of China's *Standard for Planning of Urban Green Space (GB/T51346-2019),* the service radius of mini-parks, whose size completely falls into the range of pocket parks, was set as 300 m [8].

Service proximity shows significant advantage in enhancing the accessibility of green spaces. If designed properly, pocket parks can be thoroughly integrated into residents' daily lives and serve as the main destinations for their daily recreation (Figs. 2.1 and 2.2). From the perspective of the supply side of recreational service, green spaces with high accessibility commonly have a higher level of visit frequency and service performance, which is also instrumental in the intensive use of the land resources. Meanwhile, the decentralized and proximity-oriented service mode is more favorable for precisely catering to the individualized demands of various resident groups. For example, the spatial design and facility configuration of each pocket park can be tailored to fit the exact preferences of the small groups living around, thereby enhancing the residents' degree of recreational satisfaction and well-being [9]. This trait shapes the unique values of pocket parks in optimizing the service quality and precision of urban green space system, which are also beneficial to fill the existing gap between supply and demand of outdoor recreational resources.

Fig. 2.1 Daily use pattern of residents in North-railway-street square pocket park (Nanjing, China)

Fig. 2.2 Daily use pattern of residents in Wulao-village community pocket park (Nanjing, China)

2.1.3 Segmentations

The services of pocket parks are primarily defined to fulfill the daily recreational demands of surrounding residents. However, the differences in the background of geography, climate, and culture raise discrepancies in people's daily recreation demands. According to the review of the surveys of six countries, we found that the usage patterns of pocket parks vary across different regions. Of all the surveys, the activities such as walking, fitness, socializing, pet walking, and nature contact showed relatively higher frequency. In terms of the qualities that residents concern most about in pocket parks, residents in Brisbane and Sapporo demonstrated the equivalent care on the natural and social qualities, residents in three Polish cities and Oslo of Norway showed more concerns on the natural qualities, while residents in Copenhagen and Kuala Lumpur care more about the social qualities (Table 2.2).

We also investigated the crowd activities at 74 pocket parks in the main city of Nanjing. According to the existing classification framework of recreational activities within parks and the characteristics of the outdoor recreational activities of urban residents in China [15, 16], we classified the recreational activities within the pocket parks into five types: individual physical activities, relaxation and entertainment, family and social interactions, group fitness activities, and other activities. The results showed that the proportion of people doing relaxation and entertainment (sitting, chess and card playing, eating and drinking, etc.) is the greatest, which is followed by group fitness activities (square dancing, aerobics, etc.) as well as family and social interactions (parent–child activities, neighborhood conversations, pet walking, etc.), and then by physical activities (equipment fitness, tai chi, rope skipping, shuttlecock kicking, etc.) and other activities (standing and waiting, being on the phone, etc.). This use pattern is also different from the survey results of the other six countries, as the physical activity occupies much higher rate in the park use.

As pocket parks are frequently visited by local residents, this differences in park use can be employed as a chance to observe the unique features and preferences of the local lives. Meanwhile, it also mirrors another trait of pocket park in flexible service provision.

2.1.4 Flexibility

Different from large-sized urban parks that divide its internal space sophisticatedly to support the diverse and comprehensive function system, pocket parks are seriously constrained to the spatial condition, for which their functions are usually fragmented while their activity venues are concentrated. To meet the diverse recreational demands of different groups using limited space, more efforts should be made on the enhancement of the use intensity and shareability of the space rather than space zoning or sectioning adopted in large-sized urban parks. Although this use shareability are originally the products of the passive compromise under the space constraints in most

Table 2.2 Usage characteristics of pocket parks in different cities and the qualities of most residents' concerns

Survey city	Country	Pocket park type	Most common or desired activities	The qualities of most residents' concerns
Brisbane [10]	Australia	Informal green space	Walking, enjoying the view, walking dogs, wildlife or plant observation, and natural education of children	Proximity, not crowded, more or different animals and plants, no use restrictions, and diversity
Sapporo [10]	Japan	Informal green space	Walking, enjoying the view, exposure to nature, walking dogs, and outdoor games and sports	Proximity, not crowded, more or different animals and plants, no use restrictions, and uniqueness
Krakow, Lodz, Poznan [11]	Poland	Informal green space	Walking and playing with dogs, meeting other people, viewing landscape, and experiencing wildness and nature	Microclimate, wildness, ecological importance, view, landscape, greening, and natural atmosphere
Oslo [12]	Norway	Small parks	Relaxation and philosophizing, reading/writing, eating/drinking, playing/physical activity, and socializing	A lot of grass, a lot of flowers/plants, water features, many bushes/trees, and enclosure
Copenhagen [13]	Denmark	Pocket parks	Socializing, stopover, and restitution	Cafe, other seatings than benches, tables, paved trails, and green ground cover
Kuala Lumpur [14]	Malaysia	Pocket parks	Relaxing and reducing stress, taking a shortcut, meeting people, lunch break walking, and nature accessing	–

Source Rupprecht et al. [10], Nordh and Østby [12], Pietrzyk-Kaszyńska et al. [11], Peschardt et al. [13]; Balai Kerishnan et al. [14]

occasions, it effectively activates the potential service efficiency of pocket parks, extending their spatiotemporal service capabilities objectively, and, to some extent, compensating for their congenital shortcomings on size [17].

The shared use modes of pocket parks can be categorized into two types—simultaneous sharing and time-sharing. Simultaneous sharing mainly denotes the scenario that the pocket parks accommodate diverse recreational activities of multiple groups in the same period. For example, in the 520 Student Movement Memorial Park of Nanjing, the activities such as square dancing, personal exercise, children's playing,

Fig. 2.3 The activities of 520 Student Movement Memorial Park in the evening (Nanjing, China)

pet walking, and relaxation and social interactions usually occur simultaneously at weekday nights (Fig. 2.3). Such intensive and shared use of pocket parks created many unconventional use opportunities and activity patterns. In the survey, we found that many parents or seniors who traveled with children can participate in the square dancing or other social activities, while their children can freely engage in children's playing within the observable range of their parents. In this scenario, pocket parks create convenient opportunities for both adults and children to meet their recreational demands simultaneously, whereas such usage patterns are scarcely observed in a large green space whose internal space is clearly defined by zoning.

The concept of time-sharing indicates that the use of pocket parks is dominated by various groups or recreational activities at different periods throughout the day. This phenomenon is resulted by the periodic differences in the recreational demands of different dominant users. Taking our survey on the time-sharing use mode of 74 pocket park samples in Nanjing as an example, two use transitions can be observed in the four survey periods of the weekdays. In the early morning, individual physical activities dominated the use of the pocket parks while the dominant use in the late morning and afternoon varied to the relaxation and entertainment. At night, group fitness activities became the dominant activities in the samples. On the weekends, the time-sharing use mode is basically the same as that of weekdays. The only difference is that the dominant use in the early morning shifts from individual physical activities to group fitness activities (Figs. 2.4 and 2.5).

The difference in the use pattern of different periods is also commonly seen at the individual park. The pocket park at the northwest of the intersection of Baixia Road and Qinhuai River showed significant differences in the use patterns in our four

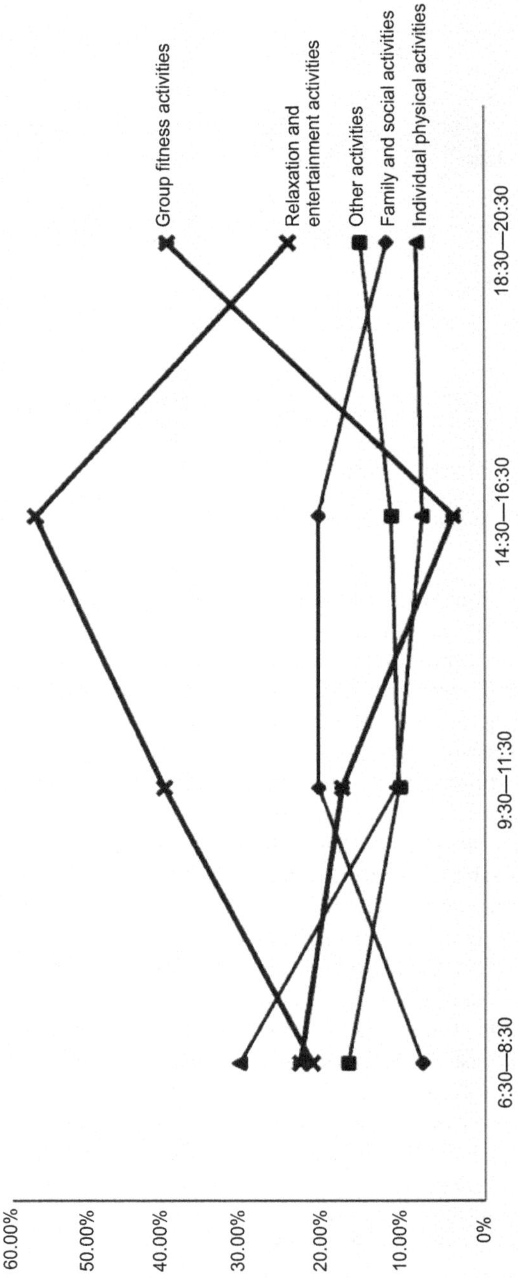

Fig. 2.4 The composition of five types of activities in four periods on weekdays

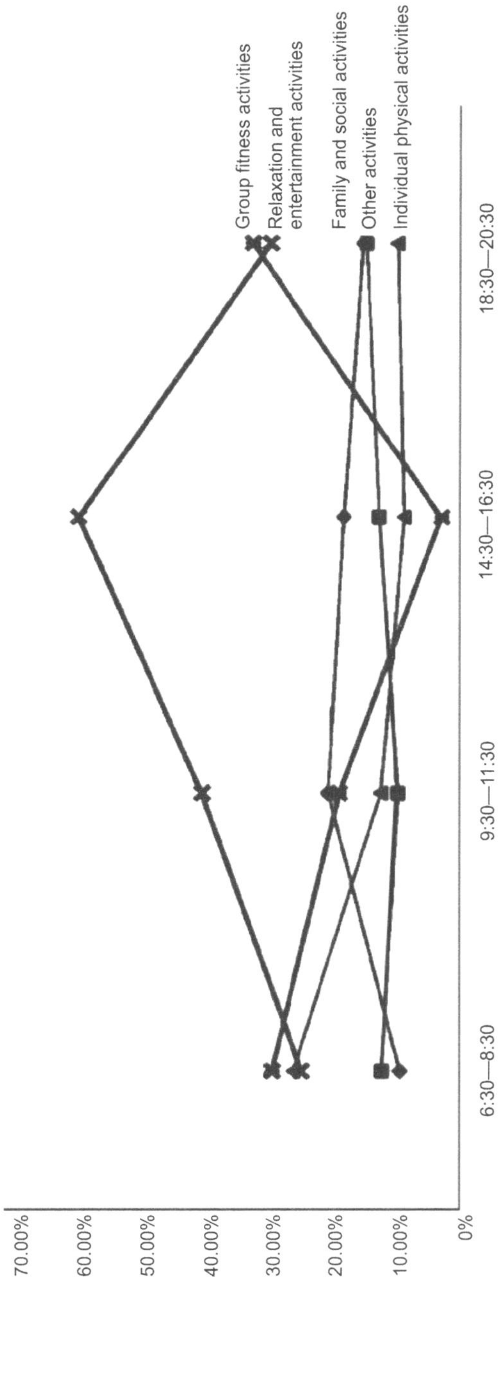

Fig. 2.5 The composition of five types of activities in four periods on weekends

survey periods, where group dancing, neighborhood communication and parent–child activities, card playing, and pet-walking dominated the park use in the early morning, late morning, afternoon, and night, respectively. The use flexibility brought by the time-sharing not only endows pocket parks with specific capability to meet the differentiated demands of different groups, but also contributes to maintain the urban vitality throughout the day, which indirectly contribute to the safety enhancement of street spaces at night (Fig. 2.6).

2.2 Spatial Attributes

2.2.1 Size and Shape

Currently, there is no unified standard defining the size of pocket parks worldwide. According to the inference of US scholar Clare Marcus, the suitable area of a pocket park is between 1 and 3 house lot units [18]. However, due to the variations in population density and land use conditions of different cities, the area range of pocket park varies. In New York City, where the available development land is constrained, the area of most pocket parks does not exceed 0.1 hm2. However, in some sparsely-populated cities of Texas, the area of pocket parks can reach to about 1 hm2 [18, 19]. Therefore, it is difficult to find an "one-size-fits-all" size standard for pocket parks. Accordingly, it is more appropriate to define pocket parks' area through relative standards according to the actual urban space and resource conditions of different regions. Currently, the relative standard with the highest consensus is that the area of pocket park should be smaller than the minimum area threshold of the local community (or neighborhood) parks [9, 20] (Table 2.3).

Besides the park size, the spatial form (size, shape, etc.) also plays an important role in park service. Seymour once suggested that the ideal size and shape of pocket parks in New York City should be 50 feet (about 15 m) × 100 feet (about 30 m) [30]; some scholars, based on the purpose of guaranteeing the space vitality and experience quality, suggest that the depth-width ratio of typical pocket parks, which are enclosed by buildings on their three sides, should not exceed 5:1 [18]. The *Public Open Space in Private Development Design and Management Guidelines* of Hong Kong not only suggested that the depth-width ratio of pocket parks should not exceed 3:1, but also proposed that the area proportion of parks' major space, commonly designed in a centralized and regular form to support most recreational activities, should exceed 75% [31]. Moreover, some other studies tried to derive parks' proper size and shape from the perspective of maintaining the intimacy of people's interaction within the parks. For example, it is found that in condition of clearly seeing the facial expressions of the talker, people can hold an effective conversation by raising their voices within a distance of 75 feet (approximately 23 m) [19].

In practice, the size control is the most popularly-used way in current standards as it is easy to be implemented and screened. The shape control undoubtedly is beneficial

Fig. 2.6 Example of use mode of time-sharing pocket parks: The pocket park at the northwest of the intersection of Baixia Road and Qinhuai River (Nanjing, China; Friday, November 13, 2020; Sunny; Average temperature 16 °C; Area 6210 m^2)

Table 2.3 Concepts related to pocket parks and their area regulations

Country or region	Year	Name	Area/hm2	Source
United States	1965	Children's playlots	0.02 ~ 0.45	Standards for Outdoor Recreational Areas [21]
		Neighborhood playgrounds	0.8 ~ 2.8	
	2020	Pocket parks	≤ 0.4	Pocket Park Toolkit [9]
The United Kingdom	2004	Pocket parks	≤ 0.4	The London Plan: Spatial Development Strategy for Greater London [6]
Australia	2008	Pocket parks	0.25 ~ 1	Western Precinct Open Space and Landscape Masterplan [22]
Japan	1995	Neighborhood parks	0.25	Urban Park Act [23]
Hong Kong, China	2011	Public green spaces	>0.05	Public Open Space in Private Development Design and Management Guidelines [24]
Chinese Mainland	1992	Residential community gardens	>0.5	Code for the Design of Public Park(CJJ 48–1992)[25]
	1993	Small parks	0.4 ~ 1	Code of Urban Residential Areas Planning & Design (GB 50,180–93) [26]
	2018	5-min pedestrian-scale neighborhood parks	>0.4	Standard for Urban Residential Area Planning & Design (GB 50,180–2018 [27]
	2018	Street garden	0.05 ~ 0.5	Technical Guidelines for the Construction of Street Garden in Shanghai (Trial) [28]
	2019	Mini park	0.1 ~ 1	Standard for Planning of Urban Green Space (GB/T 51,346–2019)
	2022	Pocket park	0.04 ~ 1	Notice of the General Office of the Ministry of Housing and Urban–Rural Development on Promoting the Construction of "Pocket Park" (Construction Office Letter [2022]No.276) [29]

to a more sophisticated guidance to park planning and design. However, the diverse perspective and principles will complicate the formulation and implementation of the control requirements, inevitably generating more cost in the whole process.

2.2.2 Constituents

Compared to larger parks, the internal composition of pocket parks is relatively simple. From the perspective of the coverage materials and use pattern, the park space can be divided into soft sites covered by greenery and water and hard sites covered by pavements, which are mainly used for scenery and activities, respectively. The early design of neighborhood playgrounds in US followed this way in spatial division and function arrangement. Its recommended prototypes placed greenery on the periphery of the site and reserved the central area for centralized recreational activities [5] (Fig. 2.7). If the site condition allows, space can also be divided based on the perspectives of activities' organized degree, dynamic and static features, and targeted groups. Additionally, this design prototype also suggests installing facilities like fitness equipment, shelters, and paddling pools to serve for the diverse groups [5].

China's *Code for the Design of Public Park (GB 51,192–2016)* stipulates that the internal space of parklands, whose area is less than 2 hectares, is mainly composed by three parts: greening (> 65%), recreational and service buildings (< 1%), and paths and paved areas (15% ~ 30%) [32]. The regulations on its facility configuration include 6 types of "required facilities" and 16 types of "settable facilities". Compared with large-sized urban parks, the requirements of the facility configuration of mini-parks, most of which can be seemed as pocket parks, are relatively simple and lenient, are much looser. The required facilities for mini-parks are all non-architectural facilities while all the architectural facilities are listed in the category of settable facilities without mandatory effect. This leaves enough discretionary space for the designers in the facility configuration based on the exact site conditions and target users' demands. (Table 2.4).

Accordingly, in the planning and design process, although pocket parks are unable to be furnished with all types of facilities and other constituents that can be commonly found in the large parks, they have greater flexibility in the selection of facilities and other constituents. In order to select the appropriate facilities and achieve to the optimal assembling of diverse constituents in pocket park design, it is necessary to break through the limitation of the traditional thinking paradigm, which pursues the comprehensive configuration of facilities and constituents. Instead, it is recommended to propose the design according to the demands of surrounding residents and form a "customized" configuration of facilities and constituents in conjunction with the site conditions. For example, due to the differences in the age group structure, residents in the newly-built apartments might prefer to setting sufficient activity venues and sports facilities for children (Fig. 2.8), while those in the old city might be more inclined to installing fitness facilities and venues for the centralized activities

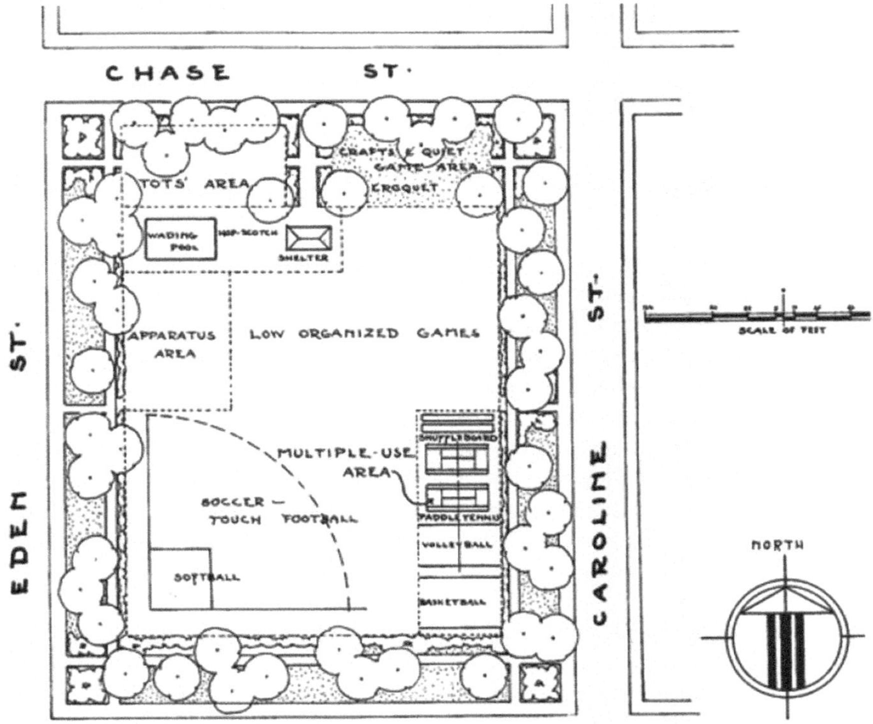

Fig. 2.7 Proposed design of typical neighborhood playgrounds: Long Range Recreation Plan (Baltimore, USA). *Source* Moeller (1965) [5]

of the seniors. This necessitates a precise response to the target users in the proposal of facility configuration and spatial arrangement. (Fig. 2.9).

In addition, for the abundant pocket parks that fall into the category of informal green spaces which are not constrained by the parkland design codes, the setting of their internal constituents can be more flexible. For example, if the regulation on soft site ratio of parks can be looser, it will create favorable conditions for enlarging the ground activity space and capacity of pocket parks. At present, the common approaches adopted in most pocket parks is to reserve as much ground space as possible for recreational activities while to maintain or increase greenery through vertical and movable greening [19, 33]. For example, if we look at the renowned Paley Park in New York City, all of its ground space, except for few tree pools, is almost reserve for activity venue. However, it is not poor of greenery on the site as various green elements can be found on the wall, movable flowerpots, and tree canopies [34]. This is also in line with the core values of pocket parks that emphasize the inclusiveness and practicality of the space.

Table 2.4 Regulations of facility configuration of mini-parks in China's *Code for the Design of Public Park (GB 51,192–2016)* [32]

Categories	Subcategories	Required facilities	Settable facilities
Recreational facilities	Non-architectural	Seats, playgrounds	Trellis, and games and fitness equipment
	Architectural	–	Pavilions, corridors, halls, and sheds
Service facilities	Non-architectural	Bike parking lots, signages, trash bins, garden lamps	Waterers, public phones, bulletin boards
	Architectural	–	Toilets, ticket offices, retail shops, medical aid stations
Management facilities	Non-architectural	Stormwater control and utilization facilities	Walls, fences, pump houses
	Architectural	–	Management and office buildings, broadcast rooms, security guarding rooms, emergency shelter facilities

Fig. 2.8 Children's Amusement facilities in 511 park (Nanjing, China)

If the constituents of pocket parks are decomposed from the perspective of three-dimensional composition, they can be categorized into four types: peripheral enclosures, ground, coverings, and focuses [35]. Of these four types of constituents,

Fig. 2.9 Seniors' Activity facilities in green space beside the Middle Longpan road (Nanjing, China)

"peripheral enclosures" include the building walls and fences that define the boundaries of pocket parks and the attached elements on them such as vertical greenery, waterfalls, and paintings; "ground" incorporates all the elements directly adhere to the ground surface, such as the ground vegetation, pavements, furniture, and amenities; "coverings" denote the elements that on the top of the park space, such as tree canopies, trellises, and hanging decorations; "focuses" serve as the focal points of the sight or activity, such as the buildings, pavilions, sculptures, and water features within the parks (Fig. 2.10).

2.2.3 Peculiarity

Pocket parks are usually partially enclosed by building walls while they also need to ensure the openness towards the street for public access. For example, the early pocket parks in New York City such as Reverend Lynette Williamson Memorial Park and Paley Park were both enclosed by building walls on their three sides and open to the street on one side [36, 37] (Figs. 2.11 and 2.12). In Hong Kong, to ensure the publicness and openness of pocket parks, it is stipulated that the street-facing openness ratio (the proportion of open boundaries in park's perimeter) of the pocket spaces should be greater than 30%, while the length of the street-facing boundary should not be less than 13.5 m [31]. The differences in urban fabric and density also produce great variations in the street-facing openness of pocket parks. For example, the street-facing openness ratio of the pocket parks in Nanjing is much higher than

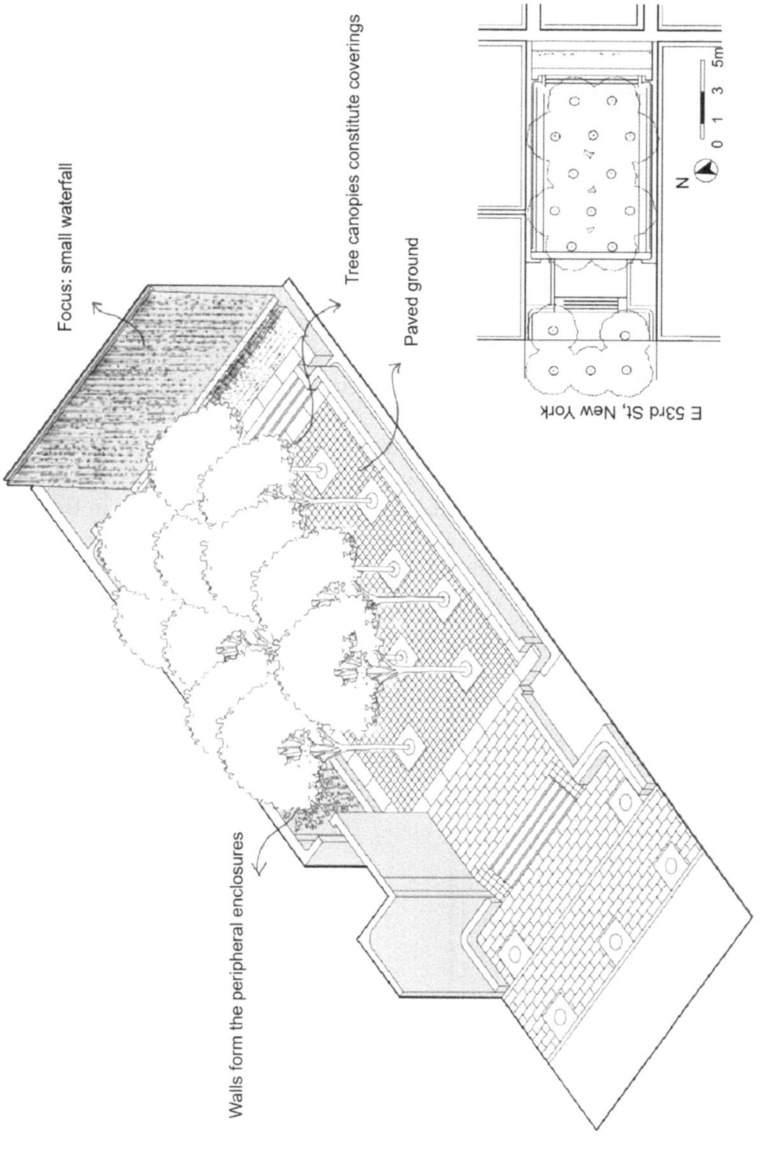

Fig. 2.10 Spatial structure and landscape elements of Paley Park (London, USA)

that in New York and Hong Kong. Of the 74 pocket park samples in our survey, the average street-facing openness ratio attained approximately 65%, indicating that the majority of the pocket parks are open to the street on two sides while this indicator in New York and Hong Kong might be much lower as most of pocket parks only have one open side (Figs. 2.13 and 2.14).

Since vast majority of pocket parks are converted from the interstitial spaces, marginal spaces, corner spaces and the like within built environments (Fig. 2.15), pocket parks are also regarded as liminal spaces within cities (Figs. 2.16, 2.17, and 2.18). "Liminality" refers to a state of in-betweenness, intermediacy, or ambiguity, namely "uncertainty in a loose space" [38]. Liminal space is usually situated at the marginal areas, characterized by emergence and flux, fluidity and malleability, and are neither segregated nor uncontained [39, 40].

L. Mahmoudi Farahani et al. totally summarized 17 related concepts (Table 2.5), which could serve as references for characterizing the spatial peculiarities of pocket parks from diverse perspectives [41]. In the service process, the fluidity and malleability of liminal space can activate more frequent interactions and connections between the parks' internal space and external environment, explaining that why pocket parks are often used as catalysts for urban renewal to infuse vitality into surrounding areas. The emergence and flux traits of the liminal space provide pocket parks with the unique diversity and dynamicity that can be hardly observed in other conventional park spaces. For example, the use surveys of pocket parks in Brisbane and Copenhagen showed that the surrounding residents took diversity as a crucial trait of the pocket parks that attracts them to visit [40]. In addition, the ambiguity and

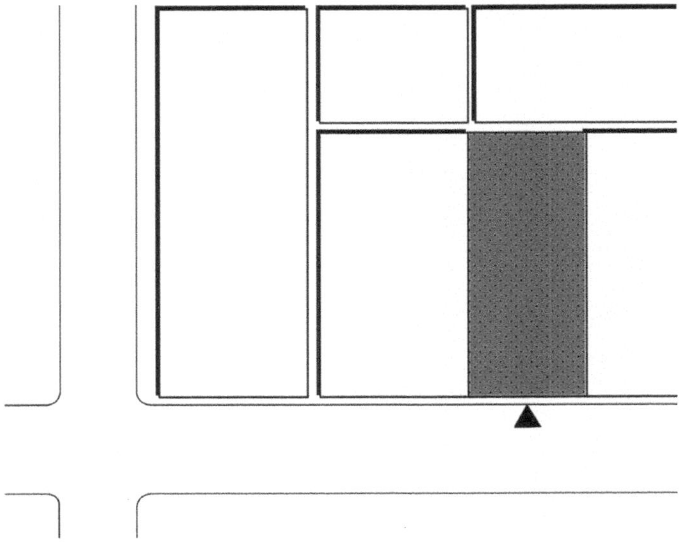

Fig. 2.11 Typical pattern of pocket parks that open to the street on parks' one side

Fig. 2.12 Image of Paley Park (London, USA). *Source* pps.org/places/paleypark

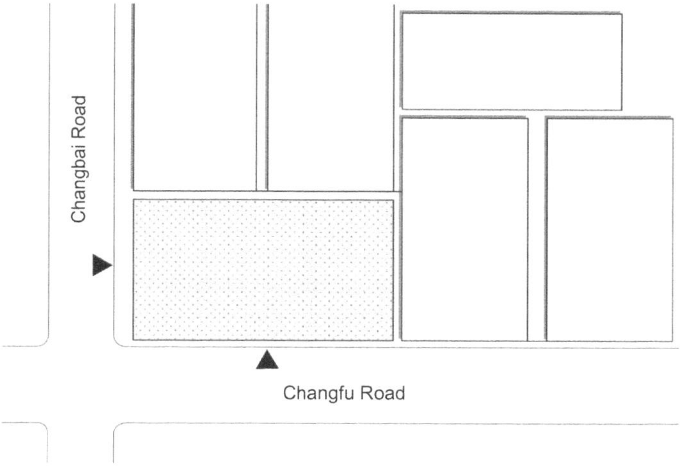

Fig. 2.13 Typical pattern of pocket parks that open to the street on parks' two side

Fig. 2.14 Reality image of pocket park near the intersection of Changbai Road and Changfu Road (Nanjing, China)

uncertainty endow pocket parks with a unique sense of mystery and explorability, which generates great attractiveness to children [42].

2.2.4 Boundaries

Unlike most large-sized urban parks occupying an independent land parcel or even an entire block, pocket parks are often located within blocks and need to share a land parcel with other facilities or buildings, for which their boundary features are much richer. The boundary of pocket parks can be classified into three categories: street-facing boundaries, building walls, and other boundaries (Fig. 2.19). The street-facing boundaries are the interface between park internal spaces and external public spaces along the streets, where most parks' entrances and exits are located. On this type of boundaries, it is easily observed the service elements like bicycle racks, sidewalk partitions, steps, greenery, and seats (Fig. 2.20). The building walls are seemed as the hardest boundary of pocket parks which is composed by building entrances and exits, building facade, and ground floor facilities (Fig. 2.21). Furthermore, apart from the aforementioned two boundaries, some pocket parks are also adjacent to the elements like rivers, municipal facilities, and so on, forming other interfaces. Such boundary forms are less common and usually non-accessible into the park space, leading to rare interaction with park's internal space.

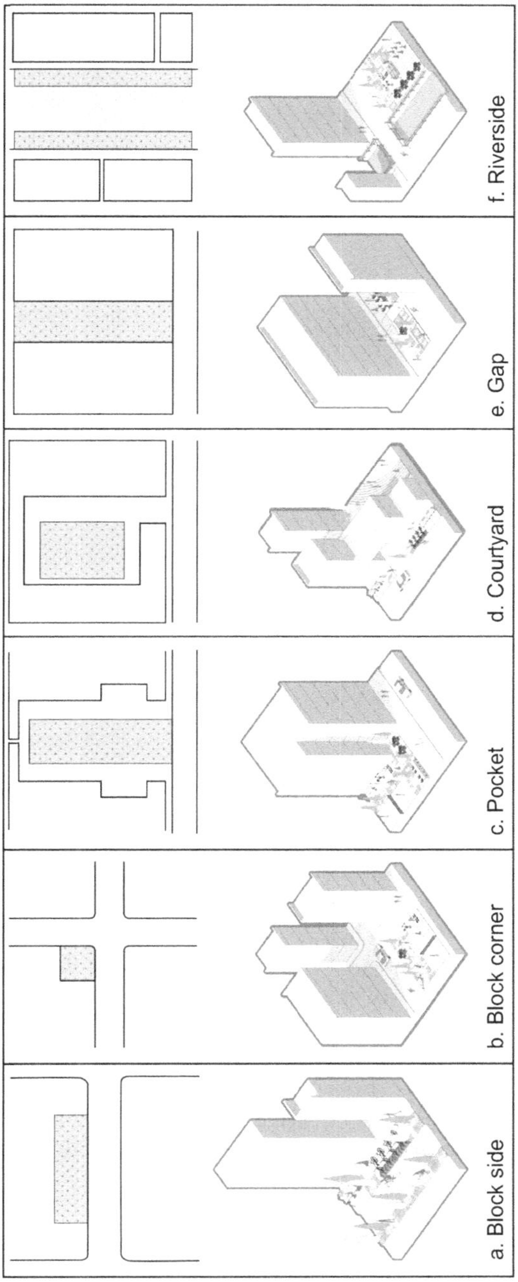

Fig. 2.15 Common positions of liminal spaces

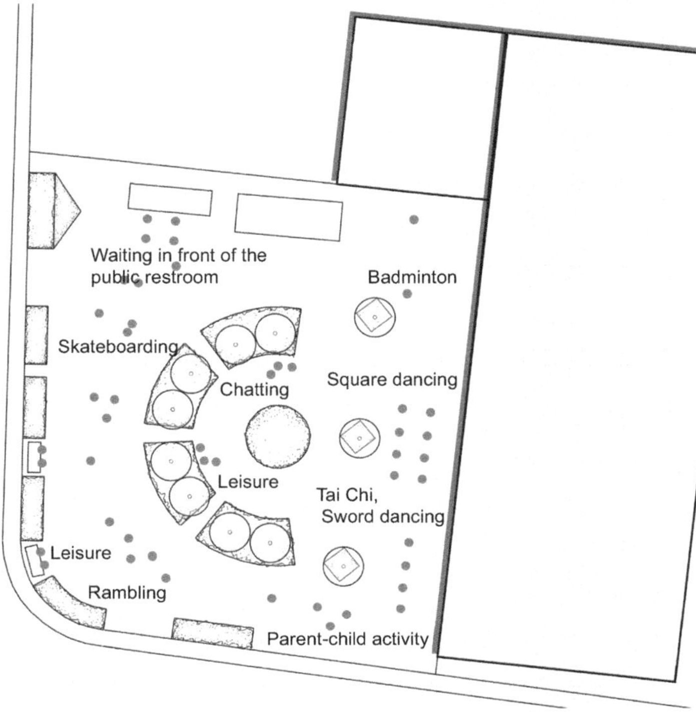

Fig. 2.16 Diversity and episodicity of activities

In the daily service of pocket parks, many boundary elements demonstrate tight correlations and interactions with the park spaces. For example, existing studies showed that the openness of the sight and traffic on the street-facing boundary directly contributes to attract more potential users to visit the parks [60, 61]. Furthermore, other studies also found that the diversity of the elements on building walls, especially the facilities in the ground floor, can help to activate the vitality and enhance the use intensity of pocket parks [18, 62]. As the forms and elements of the building walls strongly affect the service status of pocket parks, we also made a statistic on the emergence frequencies of the types of building walls based on the survey of the pocket park samples in Nanjing. It is found that the walls of the residences (36%) and offices (35%) emerged most frequently around the pocket parks, which are followed by the building walls of the finance (27%), catering (26%), education and training (24%), retail (22%), hotels (14%), shopping (14%), medical care (7%), and

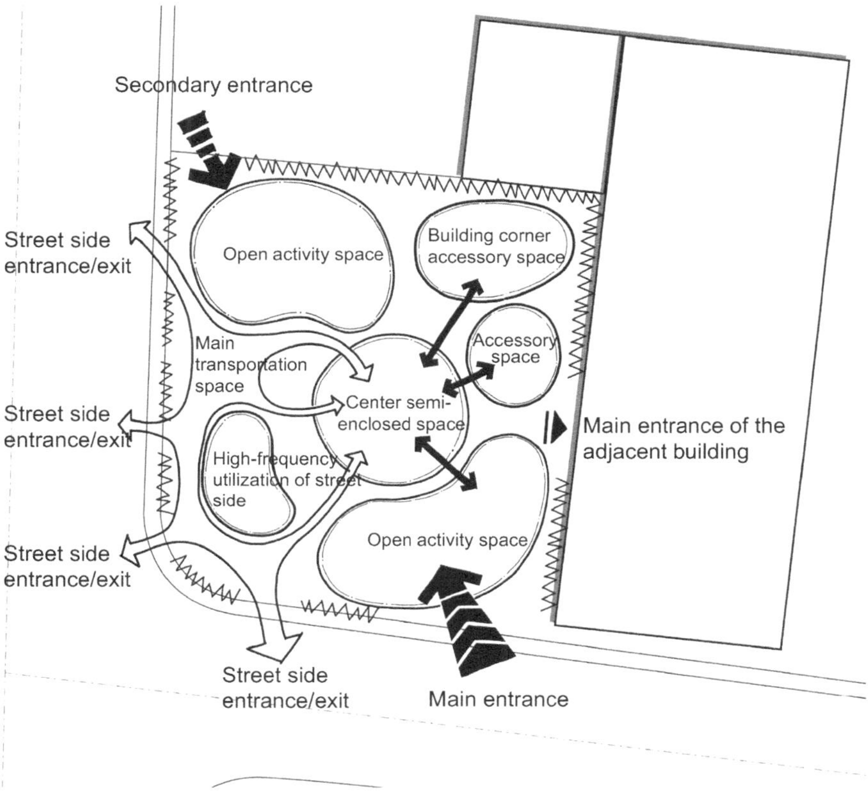

Fig. 2.17 Boundary openness and ambiguity

leisure and entertainment (7%) [Fig. 2.22]. This result also indicates the features of the frequently-developed locations of pocket parks in Nanjing.

By further observation and statistical analysis, we found that besides parks' visit frequency, the boundary facilities of buildings also directly impact their use flexibility. For example, compared to daily life service facilities (dining, retail, shopping, etc.), public service facilities (offices, financial, education, and training facilities, etc.) on the boundaries can promote the diverse use of park space more significantly. The main reason is that the front area of the entrances of the latter facilities usually reserve more concentrated and spacious sites and are well-lit during non-office hours; making them ideal places for residents to carry out individual physical activities, group dancing, neighborhood interactions, parent–child activities, and so on. However, the former facilities, as they are always in business all the time and require enough space for passage and waiting of their customers in front of their entrances, their front areas are not prone to be used for residents' recreational activities.

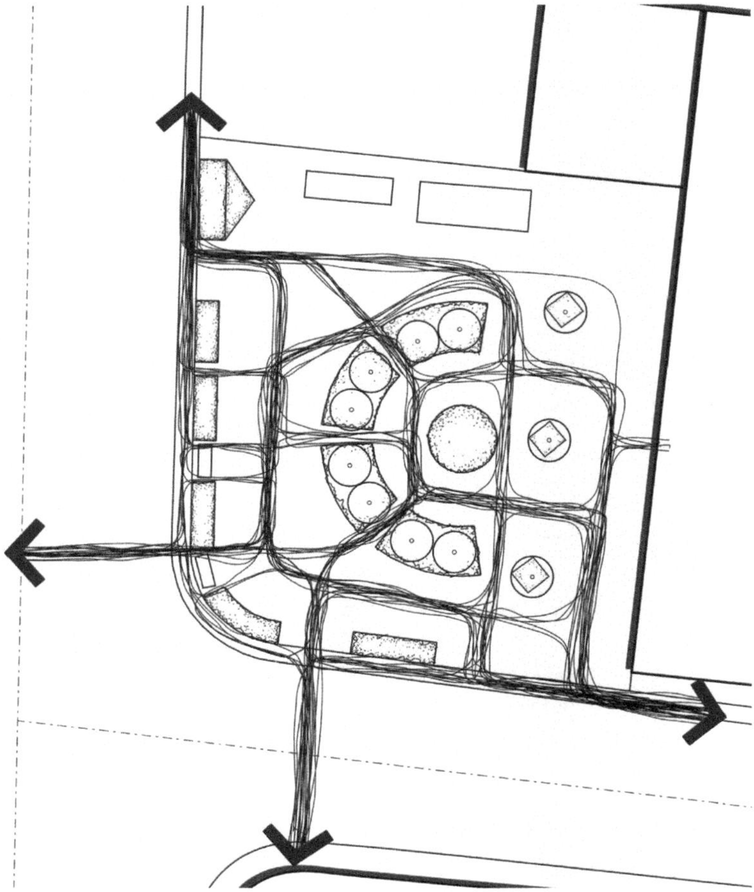

Fig. 2.18 Space fluidity

2.3 Common Forms

2.3.1 Locations

According to the positions within the blocks or streets, pocket parks can be classified into four types—namely, pocket parks at block corner (corner pockets), block side (side pockets), traversing the block (traversing pockets), and street center (street-center pockets) [18] (Table 2.6, Fig. 2.23). Of the four types, the corner pockets are relatively balanced in terms of space enclosure and openness, for which they

Table 2.5 Relevant concepts of pocket parks

Name	Source
Terrain vague	Foster [43]
Dead zones	Doron [44]
Parafunctional space	Papastergiadis [45]
Leftover space	Akkerman and Cornfeld [46]
Superfluous landscapes	Nielsen [47]
Marginal sites or wasteland	Gandy [48]
Spaces of uncertainty	Cupers and Miessen [49]
Urban voids and landscapes of contempt	Armstrong [50]
Ambivalent landscapes	Jorgensen and Tylecote [51]
Urban interstices	Tonnelat [52]
Counter public	Shaw and Hudson [53]
Unintentional landscapes	Gandy [54]
Open mosaic habitat	Maddock [55]
In-between space	Brighenti [56]
Urban wildscapes	Gobster [57]
Wild spaces	Threlfall [58]
Second hand spaces	Ziehl [59]

usually attract people's attention more easily, sharing a higher visit frequency. The side pockets are usually enclosed by building walls on multiple sides, which easily lead to a sense of closure in space and relatively weak openness towards city streets. For example, most pocket parks in New York City are side pockets [36], whose depth-to-width ratio are commonly high, leading to a relatively low usage of the distal space inside. The traversing pockets are typically in linear shape, whose opposite boundaries are open to the streets, while the other two boundaries are defined by building walls, making them frequently used as shortcut crossings and passage spaces. The street-center pockets own the highest openness as most of their boundaries are open to the public space. This unique location and spatial feature make them to be frequently used as transient spaces for passersby and decorative spaces of the city.

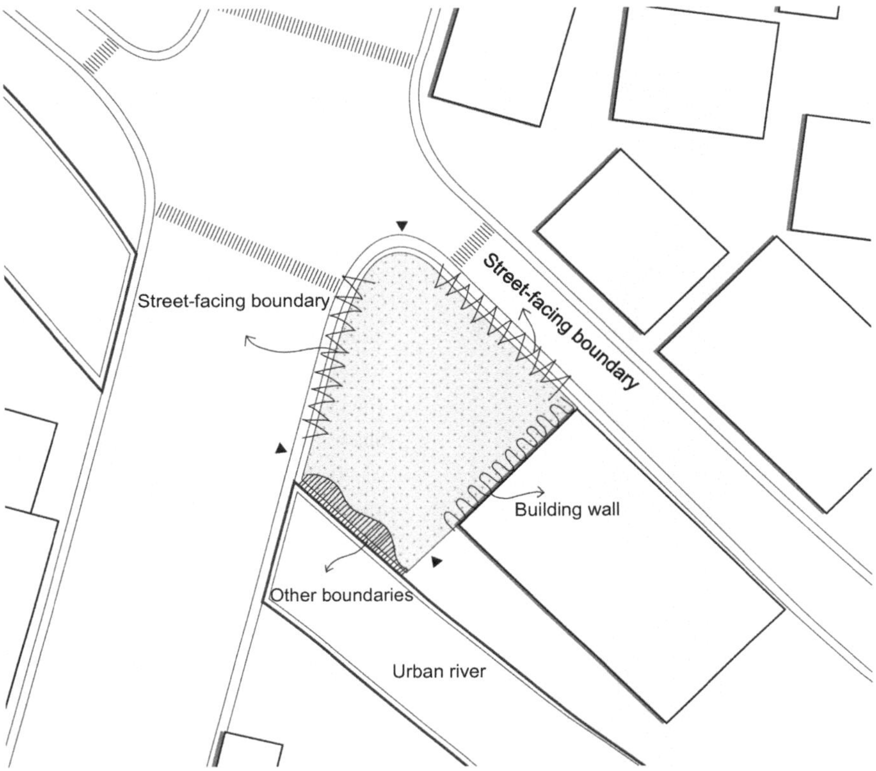

Fig. 2.19 Boundary forms of pocket parks

2.3.2 Users

Limited by space condition, pocket parks can hardly be equipped with complete functional zones and service facilities as urban parks. Instead, their functional and spatial arrangements often highlight the demands of some specific user groups. Therefore, according to the types of the dominant user groups, pocket parks can be categorized into children's playlots, youths' sports fields, middle-agers and seniors' fitness sites, and mixed-use sites. Children's play lots usually serve for the preschool children, whose facilities mainly include sandboxes, parent–child interaction facilities, plastic courts, etc. (Fig. 2.24), like the playlot (0.02 ~ 0.45 hm2) set in the *Standards for Outdoor Recreational Areas* of the United States [5]. The utilization and facility configuration of youths' sports field are more exclusive, mainly installing the courts,

(a) Sidewalk barriers

(b) Seats

(c) Stairs and seats

Fig. 2.20 Elements commonly seen at street-facing boundaries of pocket parks

tables, and other sports facilities. To avoid outside interference, they are often isolated from the surroundings through physical shields like greening, fences, and isolation net (Fig. 2.25), like the neighborhood playground (0.8 ~ 2.8 hm2) in the *Standards for Outdoor Recreational Areas* of the United States [5]. The fitness sites are mainly set to respond to the demands of middle-agers and seniors in daily socialization and fitness activities, where are commonly equipped with sufficient seats, fitness equipment, and centralized venues for group activities (Fig. 2.26). Mix-used sites, serving for diverse groups, are commonly equipped with a centralized activity area and compatible of the activities of different groups simultaneously or at different times, which are the most common ones in Chinese cities (Fig. 2.27).

(a) Building entrances (b) Building ground floor facilities

(c) Building facade elevation

Fig. 2.21 Elements commonly seen in front of building walls in pocket parks

2.3.3 Management Parties

According to the custodial subjects, pocket parks can be categorized into government-managed pocket parks, institution-managed pocket parks, and resident-managed pocket parks. Of these three categories, the government-managed pocket parks are usually formal green spaces, such as mini-parks in China's green space system, each of which independently occupies a parkland parcel. Their management and maintenance funds are mainly from the government's public financial allocation, just the same as other parklands, while their maintenance is implemented by the municipal departments of greening and parks (Fig. 2.28). The institution-managed pocket parks

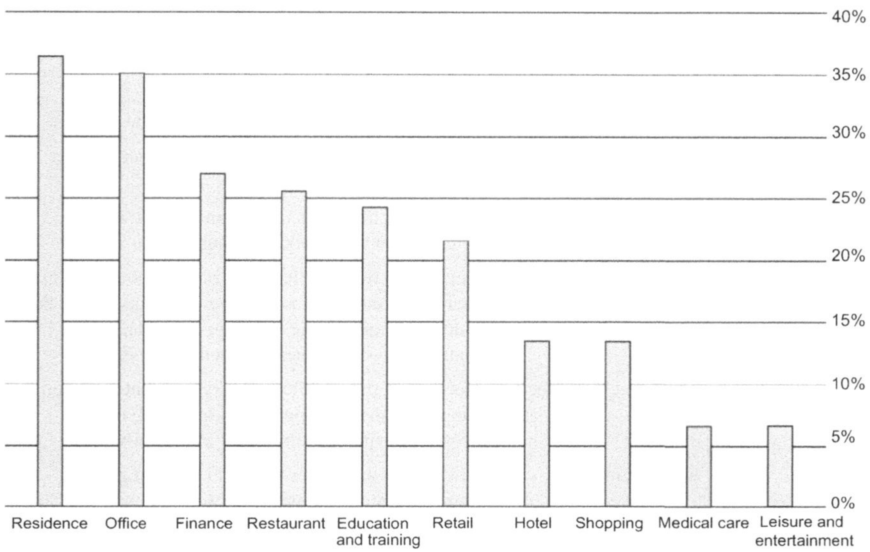

Fig. 2.22 Occurrence rate of boundary facilities of 74 pocket parks in the main city of Nanjing

are commonly in the category of the attached green spaces within the administration and public service lands, commercial and business facilities lands, and so on. In the cities of Europe and North America, these green spaces are known as "Privately Owned Public Open Spaces" (POPOS) [63]. However, in China, it is more appropriate to term these spaces as "Proprietary Public Open Spaces", such as those in the gaps between office buildings or shopping malls. Their maintenance is typically done by the owner party of the land and property such as enterprises, public agencies, and private groups (Fig. 2.29). Furthermore, resident-managed pocket parks are commonly adjacent to or within the residential areas and are the attached green spaces of residential land. This type of pocket parks can be managed by residents based on the coordination among developers, property owners, and residents' organizations (Fig. 2.30).

Table 2.6 Types and forms of pocket parks under different criterion

Criteria for differentiation	Type	Characteristics
Spatial pattern	Point shape	Serving as a recreational service center or node
	Liner shape	Providing recreational services and serving as a linkage
Spatial position	Block corner	Located in the block corner and beside road intersection with good openness
	Block side	Located in the middle of the block's boundary area, that are usually surrounded by buildings on the three sides and open to the street on one side, characterized as more enclosed and less open
	Traversing the block	Located in the middle of the block, that are open to the street on their opposite sides of the block, mostly serving for traversing and stopover
	Street center	Located in the center of the street, that are characterized as open sights, fully open to the surrounding areas, and limited access
Users	Children's playlot	Small-sized sites, mainly serving for pre-school children by children's play equipment
	Youths' sports field	Large-sized sites, mainly serving for primary and secondary school students through various types of courts and sports facilities
	Middle-agers and seniors' fitness site	Medium-sized, mainly serving for middle-agers and senior people, by fitness equipment and centralized activity venues
	Mixed-use site	Various sizes, serving for different groups
Custodial subject	Government management	Commonly formal green spaces, whose maintenance is funded by the government and implemented by the greening and gardening department
	Institution management	Commonly informal green spaces attached to public management and public services facilities and commercial services facilities, that are managed and maintained by the main owner and user of the land
	Resident management	Commonly informal green spaces attached to residential lands, that are managed by the residents on their own initiative and willingness

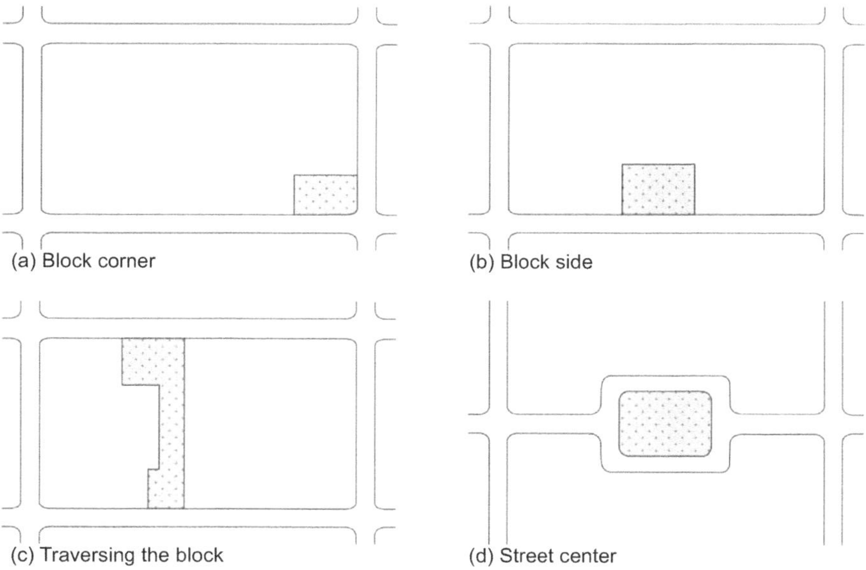

Fig. 2.23 Spatial positions of pocket parks

2.4 Comparison with Traditional Parks

As unconventional green spaces, the small size of pocket parks brings about a series of distinctive attributes in park functions and spaces. These attributes not merely spotlight the value of pocket parks in urban renewal and governance, but also mirror, to a certain extent, the limitations of their own services. To further realize and comprehend the attributes of pocket parks in planning practice, we conduct a comparison between the attributes of pocket parks and those of comprehensive parks (urban parks in western cities), which can clearly demonstrate the distinct attributes of pocket parks. It can be found that substantial differences exist between the two types of parks in terms of capacity, services, constituents, boundaries, users, and management parties (Table 2.7), which also further confirm the complementary relationship between pocket parks and conventional large parks.

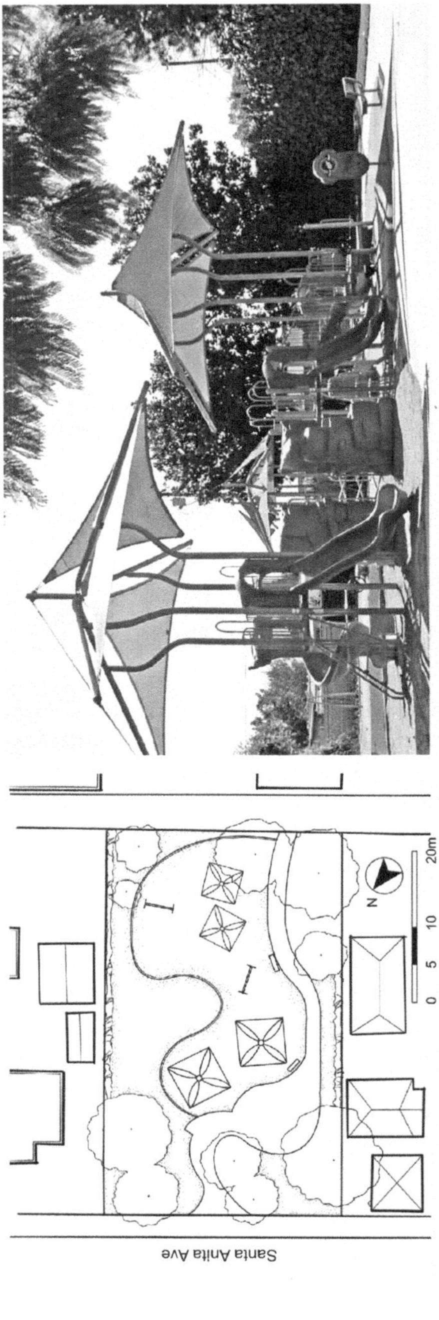

Fig. 2.24 Example of children's playlots: Plan and image of Santa Anita Playlot (Los Angeles, USA). *Image source* https://www.burbankca.gov/web/parks-recreation/santa-anita-playlot

Fig. 2.25 Example of youth's sports fields: Plan and image of Saw Mill Playground (New York, USA). *Image source* https://www.nycgovparks.org/parks/saw mill-playground

Fig. 2.26 Example of middle-agers' and seniors' fitness sites—plan and image of North Meian Pocket Park (Nanjing, China)

Fig. 2.27 Example of mix-used sited: Plan and image of Eastern Peace Park (Nanjing, China)

Fig. 2.28 Example of government-managed pocket parks: Plan and image of Jianye road Riverside Pocket park (Nanjing, China)

Fig. 2.29 Example of institution-managed pocket parks: Plan and image of Deji Plaza Pocket Park (Nanjing, China)

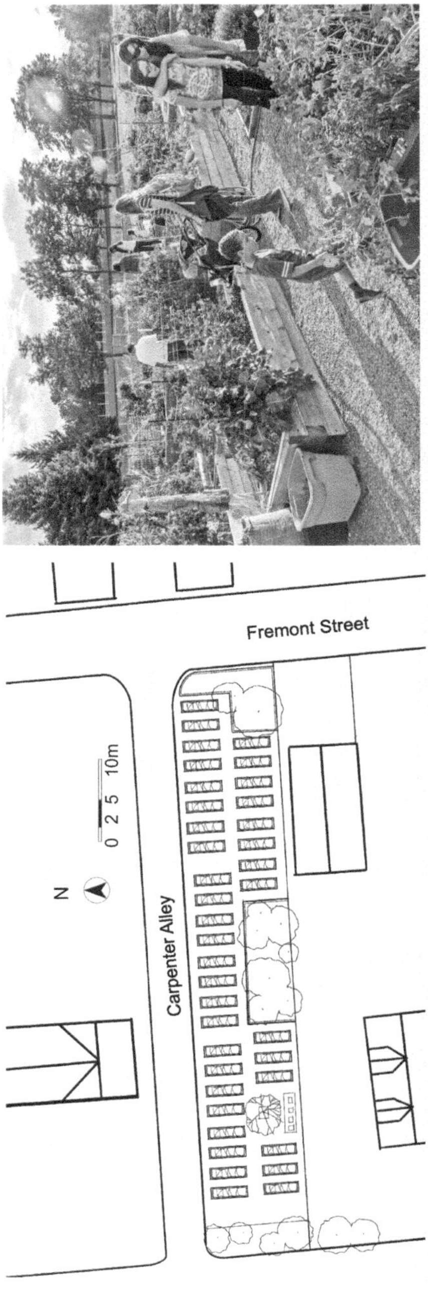

Fig. 2.30 Resident-managed pocket parks: Plan and image of St. Michaels Community Pocket Park (Liverpool, UK). *Image source* https://www.liverpoolecho.co.uk/in-your-area/community-garden-overhaul-support-food-18472210

Table 2.7 Attribute comparisons of comprehensive park and pocket park

Attributes	Comprehensive park	Pocket park
Area	>10 hm2	≤ 1 hm2
Capacity	>3 000 person	≤ 300 person
Service radius	>1 200 m	300 m
Service mode	Centralized	Decentralized
Service type	Daily and short vacation (weekend) recreation	Daily recreation
Facility configuration	Comprehensive recreational service and management facilities	Recreational facilities as the mainstay, with selective provision of other service and management facilities
Space division	Clear functional and spatial zoning	Commonly without clearly zoning and encouraging the flexible use of the internal spaces
Internal traffic	Well-established garden path system	Road system is not necessary
Position	Typically located in the center of multiple residential zones	Typically located in the corner or streetside near one community or neighborhood
Boundaries	Typically enclosed by city roads	Typically enclosed by building walls and streets
Users	Various groups of people	One specific group or multiple groups
Custodial subject	Government	Government, private institution, or local land owner and residents

References

1. Faraci P (1967) Vest pocket parks. American Society of Planning Officials, Chicago, USA
2. Zhang S, Han D, She J et al (2024) The value of pocket parks in preserving urban butterfly diversity. Urban For & Urban Green 99:128467. https://doi.org/10.1016/j.ufug.2024.128467
3. Rosso F, Pioppi B, Pisello AL (2022) Pocket parks for human-centered urban climate change resilience: microclimate field tests and multi-domain comfort analysis through portable sensing techniques and citizens' science. Energy and Buildings 260:111918. https://doi.org/10.1016/j.enbuild.2022.111918
4. Keeley M, Koburger A, Dolowitz DP et al (2013) Perspectives on the use of green infrastructure for stormwater management in Cleveland and Milwaukee. Environ Manage 51(6):1093–1108. https://doi.org/10.1007/s00267-013-0032-x
5. Moeller J (1965) Standards for outdoor recreational areas. American Society of Planning Officials, Chicago, USA
6. Greater London Authority, City Hall, The Queen's Walk et al (2008, February) The London plan-the spatial development strategy for greater London. https://www.london.gov.uk/thelondonplan.
7. Xu H (2008) A study on the system of green space planning and protection in Japan. Urban Environment Design 05:68–71

8. Ministry of Housing and Urban-Rural Development of the People's Republic of China (2019) Standard for planning of urban green space: GB/T51346–2019. China Architecture & Building Press
9. The Trust for Public Land. (2020). Pocket park toolkit. San Francisco, The trust for public land. https://www.tpl.org/resource/pocket-park-toolkit
10. Rupprecht CDD, Byrne JA, Ueda H et al (2015) "It's real, not fake like a park": residents' perception and use of informal urban green-space in Brisbane, Australia and Sapporo, Japan. Landsc Urban Plan 143:205–218. https://doi.org/10.1016/j.landurbplan.2015.07.003
11. Pietrzyk-Kaszyńska A, Czepkiewicz M, Kronenberg J (2017) Eliciting non-monetary values of formal and informal urban green spaces using public participation GIS. Landsc Urban Plan 160:85–95. https://doi.org/10.1016/j.landurbplan.2016.12.012
12. Nordh H, Østby K (2013) Pocket parks for people—a study of park design and use. Urban For & Urban Green 12(1):12–17. https://doi.org/10.1016/j.ufug.2012.11.003
13. Peschardt KK, Stigsdotter UK, Schipperijn J (2016) Identifying features of pocket parks that may be related to health promoting use. Landsc Res 41(1):79–94. https://doi.org/10.1080/01426397.2014.894006
14. Balai Kerishnan P, Maruthaveeran S, Maulan S (2020) Investigating the usability pattern and constraints of pocket parks in Kuala Lumpur Malaysia. Urban For & Urban Green 50:126647. https://doi.org/10.1016/j.ufug.2020.126647
15. Chang PJ (2020) Effects of the built and social features of urban greenways on the outdoor activity of older adults. Landsc Urban Plan 204:103929. https://doi.org/10.1016/j.landurbplan.2020.103929
16. Lachowycz K, Jones AP (2013) Towards a better understanding of the relationship between greenspace and health: development of a theoretical framework. Landsc Urban Plan 118:62–69. https://doi.org/10.1016/j.landurbplan.2012.10.012
17. Zhou C, Xie M, Zhao J et al (2022) What affects the use flexibility of pocket parks? evidence from Nanjing, China. Land 11(9):1–18
18. Marcus CC, Francis C (1990) People places: design guidelines for urban open space. New York, N.Y., Van Nostrand Reinhold. https://search.library.wisc.edu/catalog/999624093802121
19. Forsyth A, Musacchio L, Fitzgerald F (2005) Designing small parks: A manual addressing social and ecological concerns. Hoboken, N.J., J. Wiley
20. Zhou C, Zhang Y, Fu L et al (2021) Assessing mini-park installation priority for regreening planning in densely populated cities. Sustain Cities Soc 67:102716. https://doi.org/10.1016/j.scs.2021.102716
21. American Society of Planning Officials (1965) Standards for outdoor recreational areas. 1313 East 60th street—Chicago 37, Illinois. https://www.planning.org/pas/reports/report194.htm
22. St Marys Development (2008) Western precinct open space and landscape masterplan. https://www.penrithcity.nsw.gov.au/images/documents/building-development/planning-zoning/planning-controls/Western_Precinct_Plan_Part_Q.pdf
23. Japanese Government (1956) Urban park act: volume 31, section 79. https://laws.e-gov.go.jp/law/331AC0000000079
24. Development Bureau of the Government of Hong Kong (2011) Public open space in private developments design and management guidelines. Hong Kong
25. Ministry of Housing and Urban-Rural Development, PRC (1992) Code for the design of public parks: CJJ48–1992
26. Ministry of Housing and Urban-Rural Development, PRC (1993) Code of urban residential areas planning & design: GB 50180–93
27. Ministry of Housing and Urban-Rural Development of the People's Republic of China (2018) Standard for urban residential area planning and design: GB50180-2018. China Architecture & Building Press, Beijing
28. Shanghai Landscaping & City Appearance Administrative Bureau (2018) Technical guidelines for the construction of street garden in Shanghai (trial). https://www.shanghai.gov.cn/nw12344/20200813/0001-12344_55842.html

29. Ministry of Housing and Urban-Rural Development, PRC (2022) Notice of the general office of the ministry of housing and urban-rural development on promoting the construction of "pocket park" (construction office letter 〔2022〕 No. 276). https://www.gov.cn/zhengce/zhengceku/2022-08/09/content_5704766.htm
30. Seymour WN (ed) (1969) Small urban spaces: the philosophy, design, sociology, and politics of vest-pocket parks and other small urban open spaces. New York, NY, New York University Press. http://lib.ugent.be/catalog/rug01:000964963
31. Hong Kong Port Development Council (2011) Public open space in private developments design and management guidelines. Hong Kong
32. Ministry of Housing and Urban-Rural Development, PRC (2016) Code for the design of public parks: GB 51192–2016
33. Tate A (2004) Great city parks. London, Spon Press. https://search.library.wisc.edu/catalog/9910478597202121
34. Zion R (1969) Parks where the people are—the small midtown park. New York, New York University Press
35. Ashihara Y (2017) Exterior space design. Jiangsu Phoenix Literature and Art Publishing House, Nanjing
36. Armato F (2017) Pocket park: product urban design. Des J 20(sup1):S1869–S1878. https://doi.org/10.1080/14606925.2017.1352705
37. Bernstein FA (2017) Side pocket: New York's Paley Park, which turns 50 this month, is a masterpiece—proof that even a tiny public space can make a difference in a crowded city. Landscape Architecture 107(5):122–129
38. Franck K, Stevens Q (2006) Loose space: possibility and diversity in urban life (0 ed) Routledge. https://www.taylorfrancis.com/books/9781135993184. https://doi.org/10.4324/9780203799574
39. Moran D (2013) Between outside and inside? Prison visiting rooms as liminal carceral spaces. GeoJournal 78(2):339–351. https://doi.org/10.1007/s10708-011-9442-6
40. Rupprecht CDD, Byrne JA (2014) Informal urban greenspace: a typology and trilingual systematic review of its role for urban residents and trends in the literature. Urban Forestry & Urban Greening 13(4):597–611. https://doi.org/10.1016/j.ufug.2014.09.002
41. Mahmoudi Farahani L, Maller C (2019) Investigating the benefits of 'leftover' places: residents' use and perceptions of an informal greenspace in Melbourne. Urban For & Urban Green 41:292–302. https://doi.org/10.1016/j.ufug.2019.04.017
42. Chon J, Scott Shafer C (2009) Aesthetic responses to urban greenway trail environments. Landsc Res 34(1):83–104. https://doi.org/10.1080/01426390802591429
43. Foster J (2014) Hiding in plain view: vacancy and prospect in Paris' petite ceinture. Cities 40:124–132. https://doi.org/10.1016/j.cities.2013.09.002
44. Doron GM (2000) The dead zone and the architecture of transgression. City 4(2):247–263. https://doi.org/10.1080/13604810050147857
45. Papastergiadis N (2002) Traces left in cities. In: Poetics in Architecture. Wiley Academy
46. Akkerman A, Cornfeld AF (2009) Greening as an urban design metaphor: looking for the city's soul in leftover spaces. Structurist 49:30–35
47. Nielsen T (2002) The return of the excessive: superfluous landscapes. Space Cult 5(1):53–62. https://doi.org/10.1177/1206331202005001006
48. Gandy M (2013) Marginalia: aesthetics, ecology, and urban wastelands. Ann Assoc Am Geogr 103(6):1301–1316. https://doi.org/10.1080/00045608.2013.832105
49. Cupers K, Miessen M (2002) Spaces of uncertainty—Berlin revisited. De Gruyter 14–19. https://doi.org/10.1515/9783035614404-002
50. Armstrong H (2006) Time, dereliction and beauty: an argument for "landscapes of contempt." In: IFLA conference papers (pp. 116–127). The Landscape Architect
51. Jorgensen A, Tylecote M (2007) Ambivalent landscapes—wilderness in the urban interstices. Landsc Res 32(4):443–462. https://doi.org/10.1080/01426390701449802
52. Tonnelat S (2008) 'Out of frame': the (in)visible life of urban interstices—a case study in Charenton-le-Pont, Paris, France. Ethnography 9(3):291–324. https://doi.org/10.1177/1466138108094973

53. Shaw P, Hudson J (2009) The qualities of informal space: (re)appropriation within the informal, interstitial spaces of the city. In: Proceedings of the conference occupation: negotiations with constructed space (university of Brighton)
54. Gandy M (2016) Unintentional landscapes. Landsc Res 41(4):433–440. https://doi.org/10.1080/01426397.2016.1156069
55. Maddock A (2008) UK biodiversity action plan: priority habitat descriptions. JNCC, Peterborough
56. Brighenti AM (2013) Urban interstices: the aesthetics and the politics of the in-between. Routledge. http://www.tandfonline.com/doi/abs/10.1080/17535069.2015.1011432
57. Gobster P (2011) Biodiversity action plan; priority habitat descriptions. In: Urban wildscapes. Routledge
58. Threlfall CG, Kendal D (2018) The distinct ecological and social roles that wild spaces play in urban ecosystems. Urban For & Urban Green 29:348–356. https://doi.org/10.1016/j.ufug.2017.05.012
59. Ziehl M, Osswald S, Schnier D (2012) Second hand spaces: recycling sites undergoing urban transformation. Jovis, Berlin
60. Gehl J (2001) Life between buildings: using public space. The Danish Architectural Press
61. Sinou M (2013) Parameters contributing to the design of a successful urban pocket park
62. Currie MA (2017) A design framework for small parks in ultra-urban, metropolitan, suburban, and small town settings. J Urban Des 22(1):76–95. https://doi.org/10.1080/13574809.2016.1234334
63. Jiang H, Hu F, Liu W et al (2020) Research progress on privately owned open space and enlightenment to publicization of affiliated green space. Urban Dev Stud 27(11):7–11

Chapter 3
Influences on Service Performance

As outdoor recreational spaces, pocket parks' service is mainly assessed highlighting their service effectiveness and users' satisfaction [1]. The service effectiveness is commonly evaluated by visit frequency (or total number of the visitors), visit intensity, and visit duration, while users' satisfaction is usually measured through questionnaires or interviews among various user groups. Existing studies have extensively investigated park service status and its determinants, which can be categorized into internal and external factors [2–4]. Internal factors pertain to the park itself, including amenities, pathways, vegetation, water area, etc. [4, 5]. External factors relate to the surroundings of the park, and they can be divided into (1) physical factors, including surrounding land use, facilities, transportation, etc. [6], and (2) socioeconomic factors, including population size and structure, housing prices, location, etc. [7, 8].

However, existing studies mainly focused on conventional parks with large size (e.g. urban parks and community parks), whose functional and spatial attributes are significantly different from pocket parks with much smaller size [9]. For instance, some internal amenities and facilities that affect conventional parks' visits, such as pathways, water features, cafes, retail outlets, and toilets, are mostly absent in pocket parks due to their limited size [10, 11]. Furthermore, surrounding recreational facilities in existing urban contexts play a more prominent role in pocket park visits than in conventional park visits, as they compensate for the deficiencies caused by the lack of internal facilities. This also makes the service performance of pocket parks more sensitive to the variance of surroundings, calling for tailored guidelines for their planning and design in addition to those for conventional parks. Accordingly, in this chapter, we classified the results of existing studies on pocket parks' service and integrated them with the relevant findings of our team in recent years, focusing on pocket parks' service mechanism.

C. Zhou, *Principles and Methods of Pocket Park Planning and Design*, Urban Sustainability, https://doi.org/10.1007/978-981-96-5801-5_3

3.1 Internal Factors

3.1.1 Seats

Seats are one of the most common amenities in green spaces and are considered essential to maintain their vitality [12]. Although we did not find the direct evidences proving that the number of fixed seats (including benches, chairs, tree seats, etc.) significantly promote park visits in the study of pocket park samples of China, it does not mean that seats do not affect the pocket parks' service performance. For example, Karin Peschardt's investigation on the park users in Copenhagen displayed that the "variety of seats" is one of the essential attractions to local pocket park users [12]. Similarly, William Whyte observed that plazas with more sittable space tended to be more popular. Accordingly, he suggested that compared to adding fixed seats like benches and stools, it is preferable to maximize the sittability of parks' inherent features, such as ledges, pool edges, and steps [13]. Our study conducted in Nanjing further corroborates this finding, as visitors showed a preference for sittable amenities such as steps, stones, and planting edges within pocket parks, rather than fixed seats with poor location or orientation (Figs. 3.1 and 3.2).

Fig. 3.1 Sittable stones and planting edges in Nanjing Bank Building Southwest Pocket Park (Nanjing, China)

Fig. 3.2 Sittable planting edges in Dongshan Shaoye South Square Community Park (Guangzhou, China) *Source* http://way-a.com/#urbanrenewal

Meanwhile, there is another viewpoint stating that the utilization of seats in recreational areas is a result of the site attractiveness which can bring enough potential users of the seats to the site. For example, Clare Marcus found that all seats are likely to be used if the area is used intensively and attractively, indicating that intensive visits contribute to the effective utilization of seatings [14]. Similarly, Jan Gehl identified a consistent correlation between seat usage and pedestrians in his investigations of public streets in Copenhagen [15] (Figs. 3.3 and 3.4). Hence, whether seating promotes the visits of recreational spaces or visitors are initially drawn to these areas for other reasons and subsequently seek for seating is a topic debated for a long time, which may yield different conclusions contingent upon the location, characteristics, and usage scenario of the sites.

The location and layout of seats also affect the attractiveness and utilization of pocket parks [16]. Generally, seats located at the edge of a site with open views are more likely to be used than those in the center [17]. Furthermore, the seats within the sites that have various boundary features, broad views, and picturesque surroundings are more likely to be intensively used (Fig. 3.5).

For instance, our survey on the Yijiangmen-Huayengangmen segment of Nanjing City Wall Greenway proved that the tree-lined plaza equipping with the centrally-placed seatings is scarcely used. Instead, the majority of users chose to congregate

Fig. 3.3 Seat utilization in Xinjiekou Shigu Road Small Garden during peak hours (Nanjing, China)

Fig. 3.4 Seat utilization in Xinjiekou Shigu Road Small Park during junk hours (Nanjing, China)

Fig. 3.5 Center Site and its sittable edges in High School Affiliated to Nanjing Normal University East Small Garden (Nanjing, China)

on the steps situated along the southern edge of the site for relaxation and social interaction. This preference can be attributed to the panoramic view offered by the location, especially enabling parents to easily observe the activities of their children and pets within the central playground [17] (Fig. 3.6).

Compared to fixed seats, movable seats demonstrate greater popularity in pocket parks [18]. This feature arises from the diverse and variable social relationships of seat users, requiring for dynamic distance and communication patterns. Unlike fixed seats, whose inter-distance and layout cannot be adjusted, movable seats offer high adaptability to the variation of the demands from different user groups [13]. For pocket parks whose size is seriously constrained, movable seats not only fit different social distances but also facilitate a range of recreational activities, thereby enhancing the site utilization. For example, seat layout can be flexibly set as enclosure pattern to facilitate group discussion or chess (e.g., Paley Park, New York City [19]), matrix pattern to accommodate assembly activities (e.g., Bryant Park, New York City [20]), or repositioned and stacked to create broad open spaces for other concentrated activities (e.g., flea market, square dancing, etc.) (Fig. 3.7).

Movable seats are a proven method for vitalizing public spaces and indirectly contributing to the growth of surrounding commercial, retail, and other facilities, as well as city tax revenue. Jan Gehl's examination of public spaces in Copenhagen revealed that a substantial increase in movable café seats between 1989 and 1995 is positively associated with the frequency grow of recreational activities of public spaces [21]. In China, although movable seats provided by public institutions or governments are much fewer due to the maintenance and management challenges, residents in the cities like Nanjing and Shanghai were observed to frequently bring their own foldable or portable seatings to pocket parks for chess, card games,

Fig. 3.6 North site of Yijiangmen-Huayengangmen segment of Nanjing City Wall Greenway (Nanjing, China)

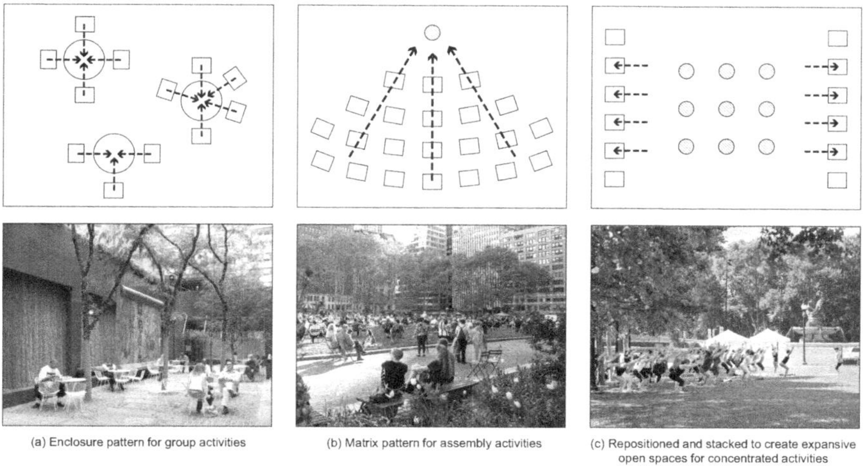

Fig. 3.7 Movable seats in pocket parks *Source* (a) https://www.ppsorg/places/paley-park (b) https://www.asla.org/2010awards/403.html (c) https://www.sasaki.com/projects/schenley-plaza/

and socializing (Fig. 3.8). This underscores the strong demand for movable seatings among pocket park users. Therefore, despite additional costs associated with maintenance and renewal, provision of movable seats within pocket parks warrants consideration as it holds promise for invigorating urban public spaces.

It should be noted that the seat itself is only one of the numerous factors that affect pocket park utilization. Except for the quantity of seats that can be easily quantified, factors like seat comfort, location, and layout are typically assessed qualitatively [22]. This qualitative nature poses challenges in accurately delineating the cause-and-effect relationships of these factors. Furthermore, seats can be assembled with other park constituents to generate joint attractiveness, such as the seats facing to the fountains or public artworks that aim for better views or seats under tree canopies that pursue for better shade. However, the impact of these factors on service performance of pocket parks can only be qualitatively analyzed at present (Figs. 3.9, 3.10 and 3.11).

3.1.2 Greenery

Greenery is a key factor in pocket park definition determining its roles to serve as green spaces. It is also the main visit reason for numerous people who are keen on a closed contact with urban nature [23]. Surveys on the pocket park utilization from various cities indicate that the most desired activities within pocket parks or the most concerned characteristics of pocket parks in people's mind are relevant with greenery, such as landscape appreciation [23], interaction with wildlife [24], diversity of flora

Fig. 3.8 Fordable or portable seats brought by residents in Pearl Hotel East Small Garden (Nanjing, China)

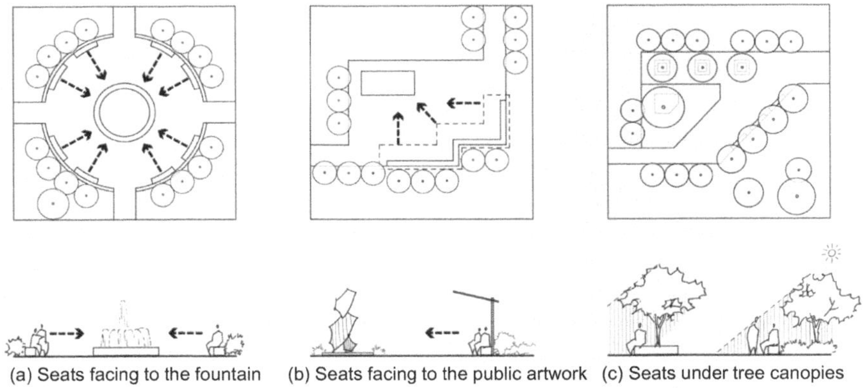

Fig. 3.9 Seats assembled with other park constituents

and fauna [25], etc. All of this confirms the essence of greenery to pocket parks. However, its impact to pocket park service performance involves a complex process.

The greenery size is a frequently discussed topic. Traditionally, it was assumed that larger greenery area or greater quantity of green space is more appealing to users. However, recent quantitative studies failed to support this viewpoint. For example, Donahue et al. (2018) used Twitter and Flickr check-in data to analyze the effect

Fig. 3.10 Seats and fountains in Floating Pocket Park (London, UK) *Source* https://merchantsquare.co.uk/sec-and-do/floatingpocket-park

Fig. 3.11 Seats and public art work in Wudaokou Center Square (Beijing, China) *Source* https://www.gooood.cn/waiting-for-the-next-ten-minutes-u-center-plaza-by-z-t-studio.htm

of tree canopy coverage on the visit frequency of various parks in the Twin Cities, MN, USA, which proved no significant correlation between the park visits and tree canopy coverage [8]. Similarly, Zhang & Zhou (2018) utilized Weibo check-in data to investigate the influence of green coverage on park visitation in Beijing, China, reaching a similar conclusion [4]. Considering that these studies did not specifically focus on pocket parks, we conducted targeted studies in the cities of Nanjing and Yancheng, China and examined the effect of the fine-grained attributes (e.g. green coverage, tree canopy coverage, lawn coverage, and shrub coverage) of greenery [26, 27]. Similarly, the results did not reveal a significant association between the greenery and park visits. All of these results jointly indicate that the existing quantitative analytical techniques may be insufficient to fully delineate this influential mechanism, highlighting the complexity of greenery's influence on pocket park service.

This mechanism also includes the influence of the greenery configuration on shaping the sense of place and other perceptual experience of pocket parks, as Jane Jacobs highlighted the significance of sightline accessibility in public spaces to its utilization efficiency and security [28]. For pocket parks where playground is centrally installed and vegetations are planted at the periphery, greenery not only defines the park space boundaries but also determines the sightlines of the pedestrian, influencing their willingness to enter (Fig. 3.12).

Before the renovation of Bryant Park of New York in the 1990s, William Whyte observed a surge in negative and criminal activities within this park. He identified that plants along the park's perimeter obstructed pedestrians' vision from the street, leading to a strong sense of unpredictability and insecurity for visitors which significantly weakened pedestrians' willingness to enter and utilize the park space [20]. Correspondingly, the core strategy raised by William Whyte is to remove these obstacles from both the perimeter and interior areas, by which an open and uncluttered visual environment was established. This simple approach greatly contributes to the revitalization of Bryant Park and makes it one of the most popular streetside parks in the city (Fig. 3.13).

In recent years, a similar approach has been adopted for the renovation of 520 Student Movement Memorial Park located in the city center of Nanjing. Previously, the park's central playground was surrounded by densely planted shrubs and trees, with only an entrance in the northwest corner. Consequently, although it occupied an ideal location, the park remained underutilized in both weekdays and weekends, contrasting with the intensely-used open spaces around. In 2020, the park underwent renovation, which involved the removal of most dense shrubs and small trees in the boundary areas, while the central area was elevated modestly to improve its visual connectivity with the surrounding street spaces. After the renovation, the site has already become one of the most attractive places for children playing, square dancing, and senior fitness in the evening for surrounding neighborhoods (Figs. 3.14, 3.15, 3.16, 3.17, 3.18, 3.19, 3.20 and 3.21).

Due to the limited space in pocket parks, excessive planting of shrubs and ground covers will inevitably compress activity areas and escalate maintenance costs. Therefore, it is advisable to avoid over-planting such greenery, except in designated zones like the segregation or buffering areas [14]. Trees with expansive canopies are

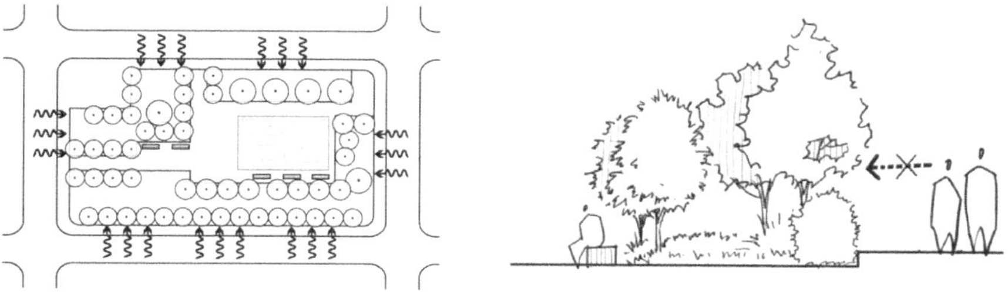

(a) Excessive greenery around the green space obstructs pedestrian sightlines and deters entry into the space.

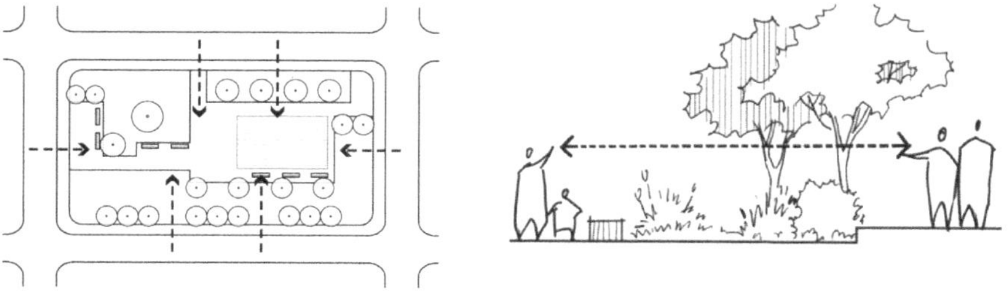

(b) Reduced greenery around the green space improves pedestrian sightlines into the space and enhances its attractiveness.

Fig. 3.12 Influence of the greenery

commonly favored because they introduce greenery and enhance the natural atmosphere without occupying the ground space. Moreover, they can improve microclimate by offering shaded areas, where are the favorite places for numerous visitors to stopover and the ideal places for installing relaxation amenities such as seats and benches. For instance, New York City raised a detailed regulation on the tree configuration in public spaces, which mandated to configure a minimum of six trees on a 5,000-square-foot (465-square-meter) plaza [13]. If abundant trees are planted in arrays, they can create interconnected and comfortable understory spaces that is conducive to recreational and communal activities for a larger user base [19] (Figs. 3.22 and 3.23).

Moreover, greenery itself can be an attraction to visitors. For instance, seasonal flowering plants can serve as attractions for both pocket parks and the entire area in the specific period. Additionally, greenery also offers habitats to various wildlife such as birds, pollinators, and small mammals like squirrels. When they are attached by informative signage about the flora and fauna, pocket parks can also serve as educational sites for local children, fostering their curiosity and spontaneous willingness to explore nature and unknown world. Furthermore, the greenery can generate more derived value if the local residents can be effectively involved in the planting and

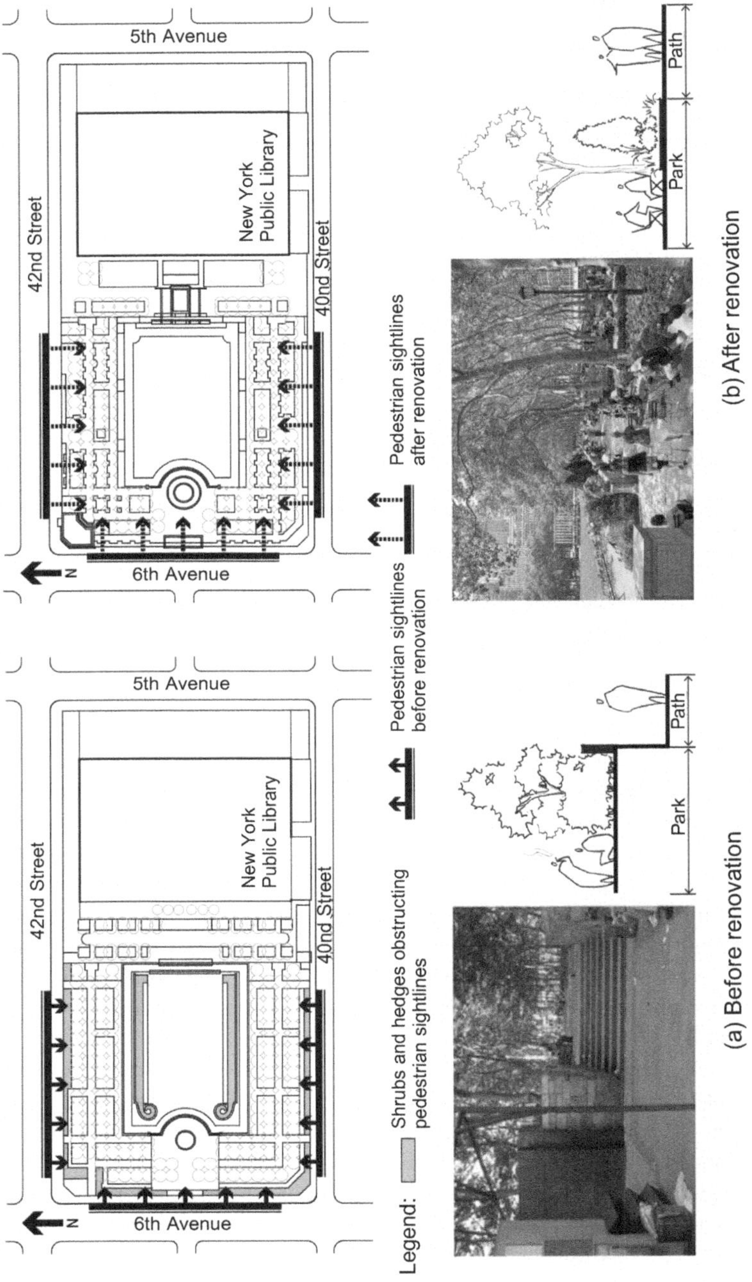

Fig. 3.13 Renovation of Bryant Park (New York, USA) *Source* https://www.asla.org/2010awards/403.htm

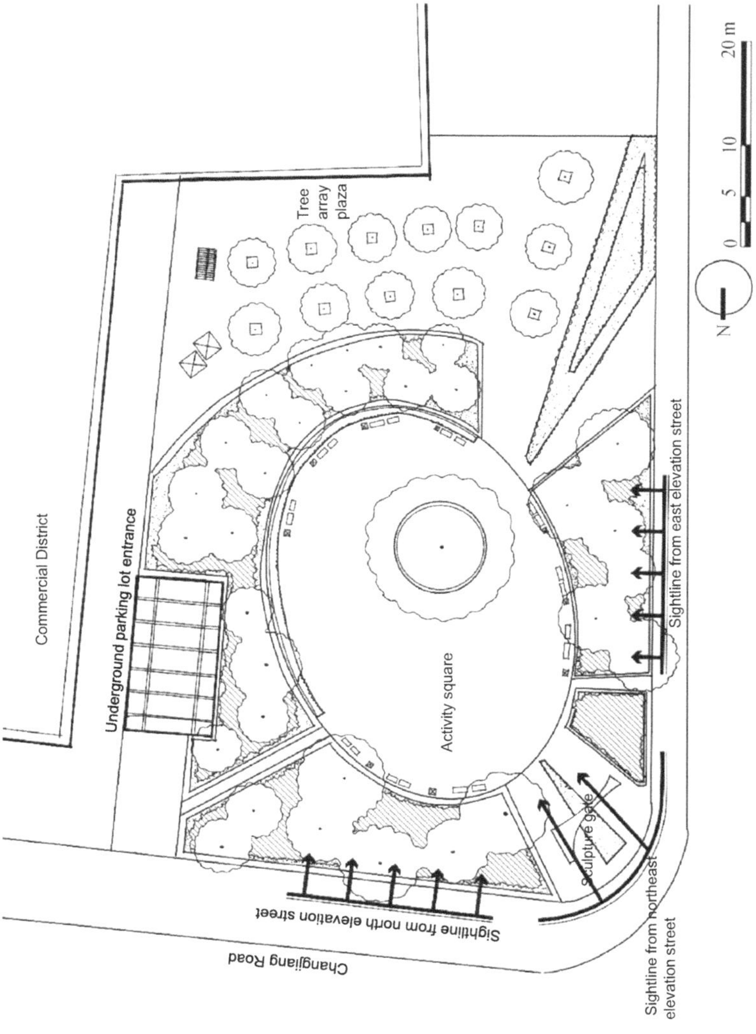

Fig. 3.14 Plan of 520 Student Movement Memorial Park before renovation (Nanjing, China)

Fig. 3.15 Northeast site before renovation

Fig. 3.16 North boundary before renovation

Fig. 3.17 East boundary before renovation *Source* Baidu Street View, https://map.baidu.com/

maintenance of the parks. A notable example is the Lafayette Greens in Detroit, USA, which won the 2012 ASLA Honor Award. The designers and local residents jointly transformed a 1,700 square-meter abandoned site in the downtown into a vibrant community farm, which not only provides access to nature but also facilitates parent–child activities, neighborhood communication, education, and the sharing of harvests [29] (Figs. 3.24, 3.25 and 3.26). Similarly, the Chuangzhi Farm in Shanghai, China, successfully revitalized an unusable land, fostering intimate interaction and communication among residents (Figs. 3.27 and 3.28). However, since residents' participation in greenery maintenance is easy to trigger the issues related to the responsibility subject, security guarantees, technical support, and dispute coordination, most of current explorations mainly focus on informal green spaces where the managements and constraints are much looser and the barriers in implementation are less than formal parks.

3.1.3 Playground

If we review the core value of pocket parks, it is easy to reach a conclusion that the space utilization is far more important than the space viewing [27]. The playground is the main carrier for most recreational activities within pocket parks. According to China's *Code for the Design of Public Park (GB 51,192–2016)* [30], the installation of facilities within parks should prioritize the playground [26, 31]. The main

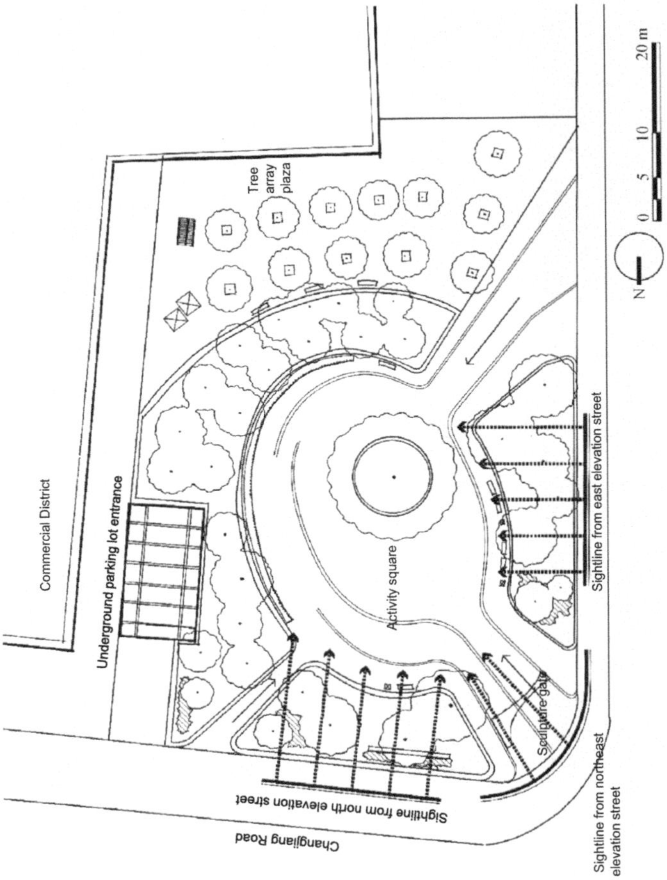

Fig. 3.18 Plan of 520 Student Movement Memorial Park after renovation (Nanjing, China)

Fig. 3.19 Northeast site after renovation

Fig. 3.20 North boundary after renovation

Fig. 3.21 East boundary after renovation

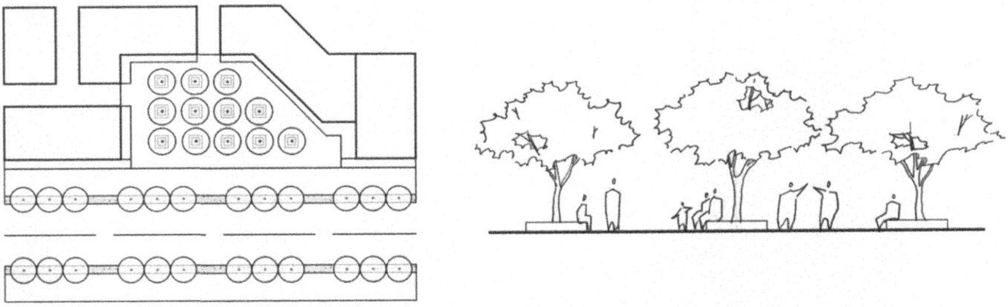

Fig. 3.22 Tree array and the under-canopy space in pocket parks

reason is that the area of playground directly determines park capacity as well as the types of available activities. This role is particularly pronounced in densely populated areas with limited recreational space [12].

Hamstead et al. (2018) found that the presence of centralized playground exceeding 100 m^2 can notably enhance park visit frequency across a variety of parks according to the survey of New York [32]. Our findings in Nanjing and Yancheng also revealed a significant positive association between tourist numbers and the size of centralized playground [27, 33]. The effect of this variable on the variance of the park visits is even stronger than that of the park's total size (Fig. 3.29). This underscores that the primary determinant of pocket parks' capacity lies not in the park size but in the area of their playground. Therefore, in the built-up area where the

Fig. 3.23 Jianye Road Riverside Mini-Park (Nanjing, China)

area available for pocket park development is strictly limited while the surrounding recreational demands are intensive, ensuring the area of playground or its percentage in park's total area can enhance the response degree to the recreation demands. For instance, the Public Open Space in Private Development Design and Management Guidelines of Hong Kong requires that at least 75% of the total area of pocket parks should be designated as main playground for intensive activities [34].

The spatial quality and user experience of the playground also sophisticatedly influence the service performance of pocket parks. William Whyte argued that "*Unless there is a compelling reason, an open space shouldn't be sunk*", because sunk space makes the people inside feel like being "*at the bottom of a well,*" where the visibility and interaction with the surroundings are restricted [13]. The San Francisco Planning Department has conducted a survey on the utilization of open spaces along Market Street, finding that the use intensity of the sunken sites is much lower than that of the nearby non-sunken sites [35]. Similarly, Hong Kong government also requires that pocket spaces should be "clearly" visible [34]. Our research in the pocket parks of Nanjing yielded comparable findings. For example, the sunken site in front of Lizhi Square Building saw minimal use during peak hours despite its large size, superior surrounding greenery, and lighting, while the adjacent non-sunken Nanjing Bank Building Southwest Pocket Park is bustling with activity (Fig. 3.30).

The utilization of pocket parks can be impacted by elevation variations, as height differences lead to spatial division and constrain the flexible use of playground. For example, group activities such as square dancing and aerobics, which require ample space, may not be available to the sites with varying heights (Figs. 3.31 and 3.32). Specifically, the dense variation of the ground heights within a small area will prevent the flexible use of park spaces and erase the diversity of park utilization, finally impacting the service efficiency of pocket parks. To avoid this issue, Hong

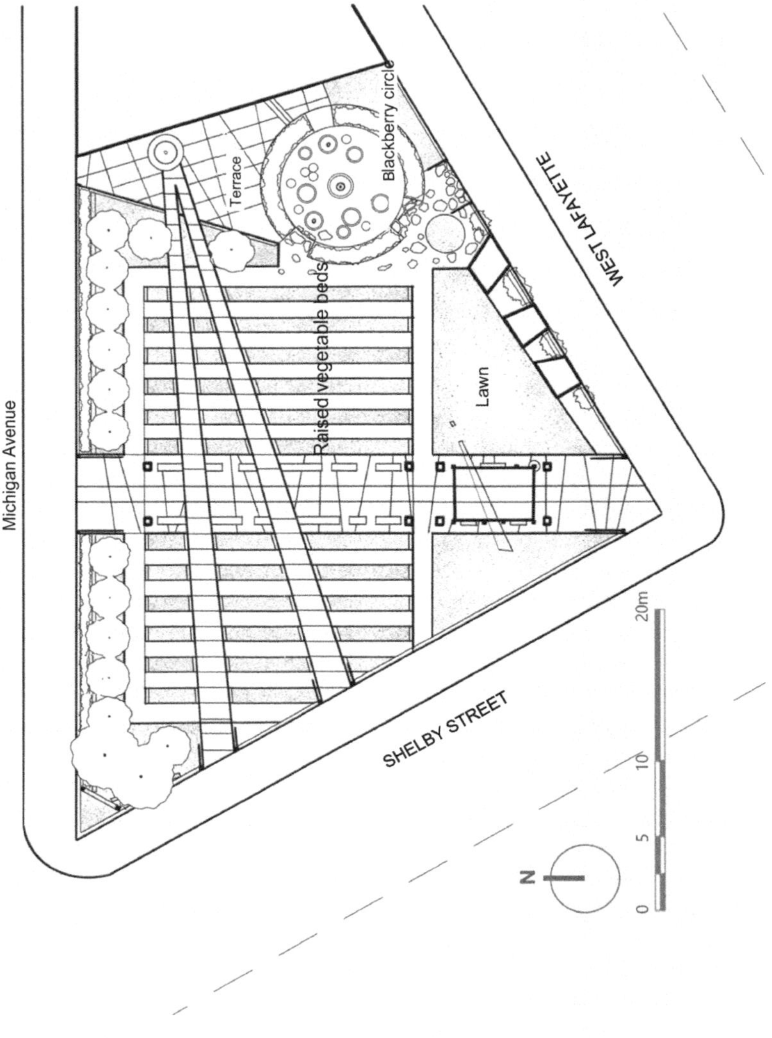

Fig. 3.24 Site plan of Lafayette Greens (Detroit, USA)

Fig. 3.25 Aerial view of Lafayette Greens (Detroit, USA) *Source* https://www.asla.org/2012awards/073.html

Fig. 3.26 Vegetable planting area in Lafayette Greens (Detroit, USA) *Source* https://www.asla.org/2012awards/073.html

Fig. 3.27 Chuangzhi Garden (Shanghai, China) *Source* http://k.sina.com.cn/articlc2138152110c bc8fab02000wcc7.htm

Fig. 3.28 Residents' participation in in vegetable planting within Chuangzhi Garden (Shanghai, China) *Source* https://www.jianshu.com/p/54cf5a7d277f

Fig. 3.29 Influence of the size of centralized playground on park visits Pocket Park A. Huowa Lane East Pocket Park (Nanjing, China); Pocket Park B. Fuqiao East Mini-Park (Nanjing, China); Pocket Park C. Song Ziwen House South Mini-Park (Nanjing, China)

Kong recommends to prioritize to install small green spaces on flat terrain that are free of obstructions to encourage the occurrence of a variety of group activities [34].

In our survey, we observed the shape diversity of the playground within pocket parks. However, in condition that the playground allows for concentrated activities, the shape of the site does not significantly affect its internal activities or the use flexibility (Figs. 3.33, 3.34 and 3.35). According to Karin Peschardt's study on pocket parks of Copenhagen, Denmark, it is proposed that the centralized playground should not be smaller than 20 m × 20 m, aiming to maintain the activity compatibility of the park space [12]. However, considering the higher population density of Chinese cities and the compactness of their public spaces, the centralized playground in Chinese pocket parks can be relatively smaller. In our investigation, a size of larger than 12 m × 12 m is observed to be capable of meeting most daily recreational demands of residents in Chinese cities (Fig. 3.36).

Fig. 3.30 Plan and evening utilization of the sunken site in front of Lizhi Square Building and the non-sunken site of Nanjing Bank Building Southwest Pocket Park (Nanjing, China): (a) Plan of the sunken site in front of Lizhi Square Building; (b) Plan of the located neighbourhood of the parks; (c) Plan of Nanjing Bank Building Southwest Pocket Park; (d) Evening photo of the sunken site in front of Lizhi Square Building; (e) Evening photo of Nanjing Bank Building Southwest Pocket Park

3.1.4 Other Amenities

Although pocket park cannot be equipped with an extensive range of amenities similar to larger parks due to the limited space, installing amenities such as public restrooms, small cafes, and kiosks can offer great convenience to park users [36]. Moreover, smaller non-building amenities can enhance the pocket park quality and attractiveness. For example, drinking fountains play a crucial role in catering to various park users, especially for children who need to drink regularly. These fountains not only attract more visitors but also encourage longer stays, particularly in hot summer or for the pocket parks where are defined as the main destinations for sports and fitness activities [14] (Fig. 3.37).

Multi-functional tables and chairs serve as valuable assets for facilitating social interactions among park users [37] (Fig. 3.38). Unlike traditional linear benches that inhibit face-to-face communication, multi-functional tables and chairs can effectively

Fig. 3.31 Kunlun Road Community East Mini-Park (Nanjing, China)

Fig. 3.32 Shuiximen Square West Mini-Park (Nanjing, China)

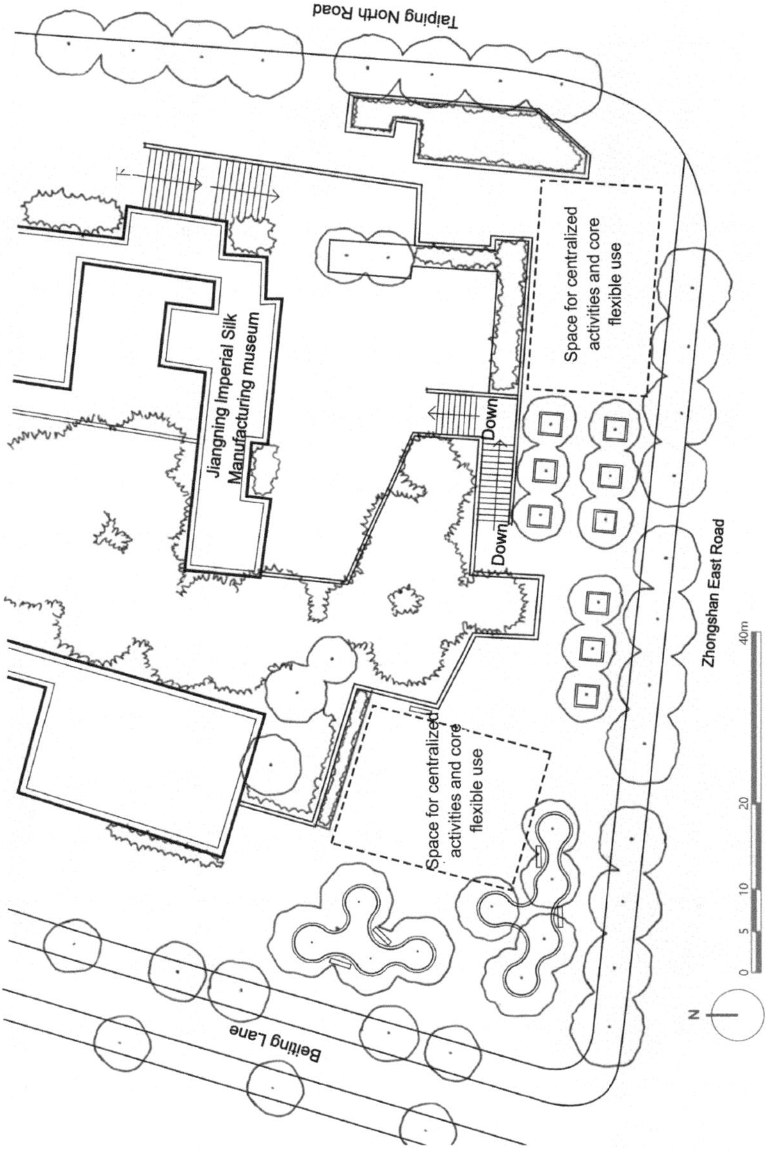

Fig. 3.33 Internal playground of Jiangning Imperial Silk Manufacturing Museum South Green Space (Nanjing, China)

Fig. 3.34 Recreational activities in playground of Jiangning Imperial Silk Manufacturing Museum South Green Space (Nanjing, China)

Fig. 3.35 Group activities in playground of Jiangning Imperial Silk Manufacturing Museum South Green Space (Nanjing, China)

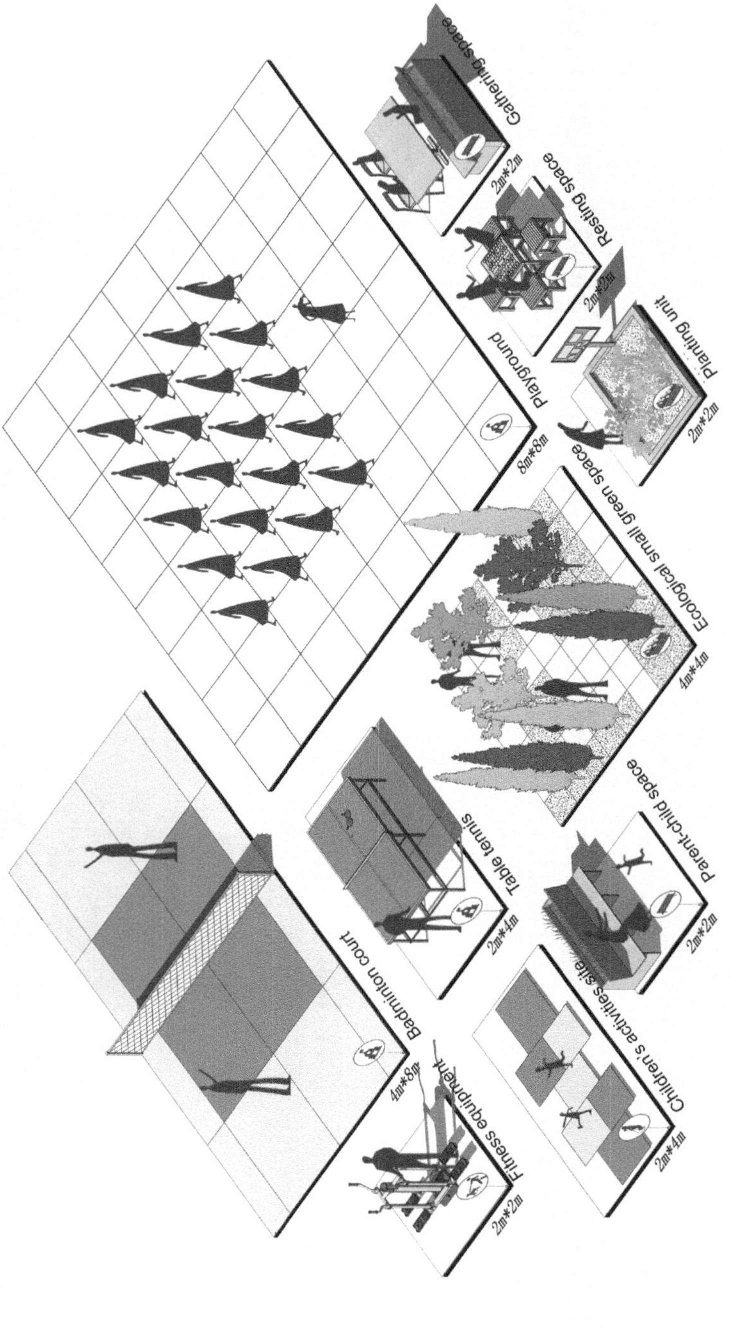

Fig. 3.36 Typical recreational activities in playground

Fig. 3.37 Drinking fountains in An'ning People's Hospital East Mini-Park (Kunming, China)

create one-on-one or group communication spaces to support activities like chess, card games, and outdoor dining, compensating for the deficiencies of benches and leaving more use flexibility to park users.

Well lighting is also considered essential for guaranteeing the safety and usability of pocket parks during the peak hours in the evening and night, particularly for the activities such as square dancing, pet gatherings, and children's playing [36] (Fig. 3.39). Additionally, moderate lighting enhances visibility within the park, promoting safety for nighttime park users and passersby (Fig. 3.40).

3.2 External Physical Factors

3.2.1 Recreation-Related Facilities

The absence of a comprehensive system of recreation facilities in pocket parks directly affect the integrity of their recreation services [38]. However, pocket parks are usually located in high-density environments, and the existing recreation-related facilities in their boundaries or surrounding areas can make up for the inherent deficiencies of their services. Therefore, if the pocket park is located in an area where

Fig. 3.38 Multi-functional tables and chairs in the Republic Public Realm (London, UK) *Source* https://www.gooood.cn/republic-public-realm-studio-rhe.htm

Fig. 3.39 Nanjing General Labor Union Building West Pocket Park (Nanjing, China)

Fig. 3.40 Nighttime lighting in Mei 'an West Pocket Park (Nanjing, China)

recreation-related facilities are densely distributed, it will activate a positive interaction and complementary effect between the pocket park and surrounding facilities, which would not only improve the quality of the park's recreational services but also increase the vitality and economic return of the surrounding facilities [15] (Fig. 3.41). According to the recommended facility categories listed in China's *Code for the Design of Public Parks (GB 51,192–2016)* [30], the probable absent facilities in pocket parks that are closely related to daily recreation include toilets, catering, and retail shops. Through a corresponding investigation and analysis, we found that these three types of facilities located in the surrounding area can also support the service of pocket parks, although they might act in different ways.

The public toilets in the surroundings can effectively compensate for the service deficiencies caused by the lack of such facilities in the pocket park. They are also directly related to park service for the long-time activities, such as square dancing, children playing, card and chess playing, etc. Studies showed that children commonly need to use toilet after an hour of activity in pocket parks, and seniors use toilet more frequently. Our analysis in the central city of Yancheng during the peak hours of outdoor recreation (17:00–19:00) demonstrated that the installation of public toilets around the pocket parks significantly and positively associates with park visits [27]. It is important to note that not all the surroundings of pocket parks are equipped with independent public toilets due to realistic constraints. Toilets inside public buildings, such as shopping malls, restaurants (e.g., KFC, McDonald's, etc.), and large public facilities (e.g., libraries, fitness centers, etc.), can also serve the visitors of nearby pocket parks (Fig. 3.42) with ease.

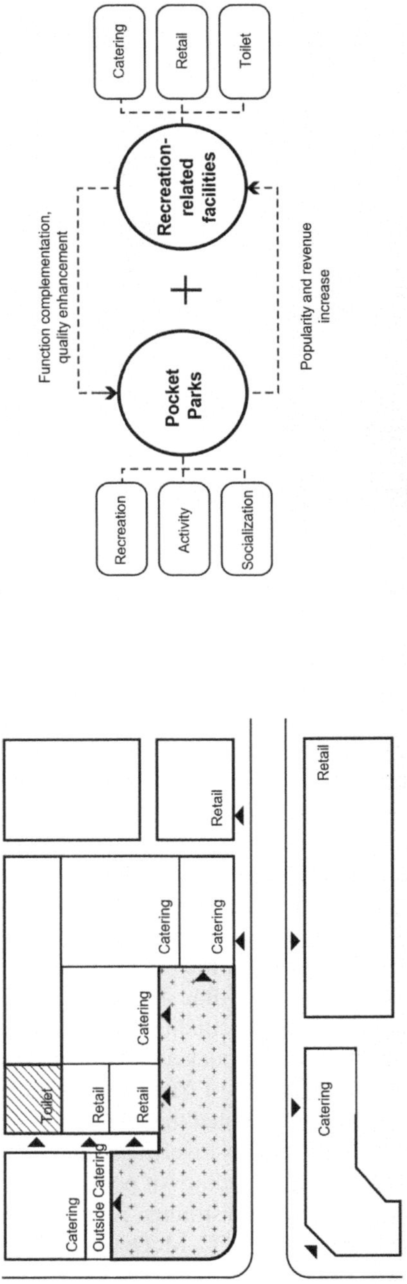

Fig. 3.41 Interaction and mutually complementary effect of "Pocket parks + Recreation-related facilities"

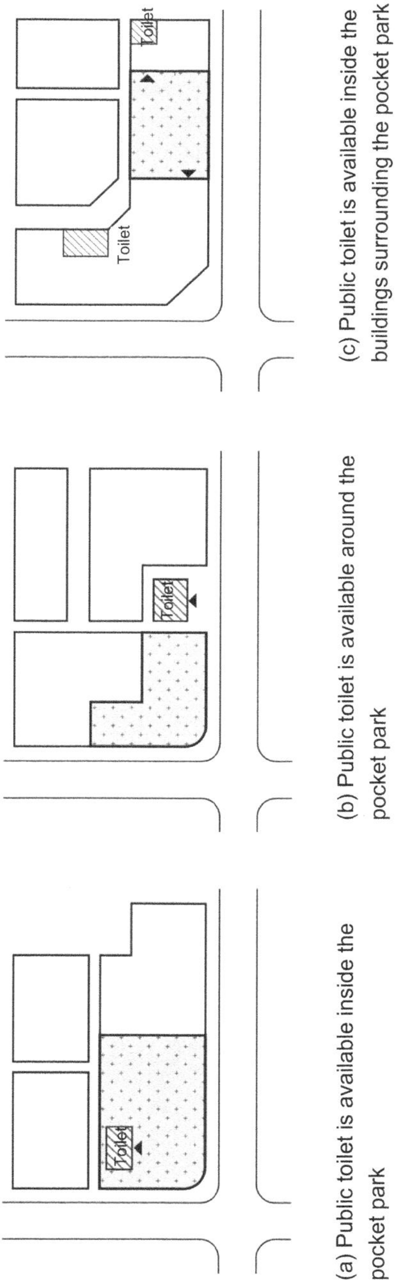

Fig. 3.42 Installation of public toilets around pocket parks

The surrounding catering facilities not only provide a scenic dinning place for park visitors but also create additional spaces for their socializing. For example, Peschardt et al. (2016) observed that cafes effectively foster social interaction among users and become a highly favored attribute of pocket parks in Copenhagen [12]. Likewise, Gehl (1996) indicated that cafes with outdoor seats contribute significantly to the vibrancy of adjacent public spaces by enhancing social activities [21]. Similarly, in our quantitative study in Nanjing, we found that the proximity of catering facilities positively influences pocket park utilization. This proximity not only makes up the parks' service limitations, but also brings a considerable number of potential users to the parks. Moreover, catering facilities also benefit from the presence of pocket parks. For example, existing studies proved that the abundant park visitors have great potential to be converted into customers for these facilities, thereby boosting their turnover and contributing to city's tax revenue (Figs. 3.43 and 3.44).

For those visitors who prefer not to utilize fixed catering facilities, surrounding retail shops such as kiosks and convenience stores can offer them portable food and beverage to be enjoyed within parks (Fig. 3.45). The nighttime lighting of the shopwindows of surrounding stores can also enhance the vitality and sense of security, guaranteeing parks' use efficiency and stability [21] (Fig. 3.46). However, compared to toilets and catering facilities, surrounding retail shops showed weaker effect on promoting the utilization of pocket parks. This partly attributes to the less necessity of the retail services to park visitors, especially the majority visitors of daily recreation. Moreover, the functional overlap in food and beverage sales between retail shops and

Fig. 3.43 Catering facilities on the periphery of Schenley Plaza (Pennsylvania, USA) *Source* https://www.sasakicom/projects/schenley-plaza/

Fig. 3.44 Catering facilities on the periphery of Schenley Plaza (Pennsylvania, USA) *Source* https://www.sasakicom/projects/schenley-plaza/

catering facilities is probably the other reason, leading to the mutual substitutability of their service.

3.2.2 Transportation Systems and Related Facilities

The location, structure, and forms of the slow-traffic system around pocket parks directly determine the access modes and effectiveness of potential users of pocket parks, thereby impacting parks' accessibility and connectivity with surroundings. The proximate service mode of pocket parks results that the primary means of surrounding residents to access parks are walking and cycling [39]. Therefore, a well-designed and -maintained slow lane that links the pocket park with local residences can bring great convenience for local people to obtain park service. Similarly, effective linkage between surrounding commerce or offices and their surrounding pocket parks can introduce abundant potential park users during lunch break, off-work, and peak shopping hours. Moreover, high connectivity between pocket parks and other forms of public spaces in the vicinity can promote the integration of the pocket park service into the regional recreational system, thereby increasing the probabilities of park visit [33] (Fig. 3.47).

In addition to transportation connection, if the pocket park is open enough to the street space or adjacent slow lane, the active and busy traffic corridor can also

Fig. 3.45 Visitors enjoying portable food and beverage in Greenacre Park (New York, USA) *Source* https://www.sasakicom/projects/greenacre-park/

Fig. 3.46 The nighttime lighting of stores around Zuccotti Park (New York, USA) *Source* https://commons.wikimedia.org/wiki/file:ZuccottiParkatnightoctober_2014.jpg

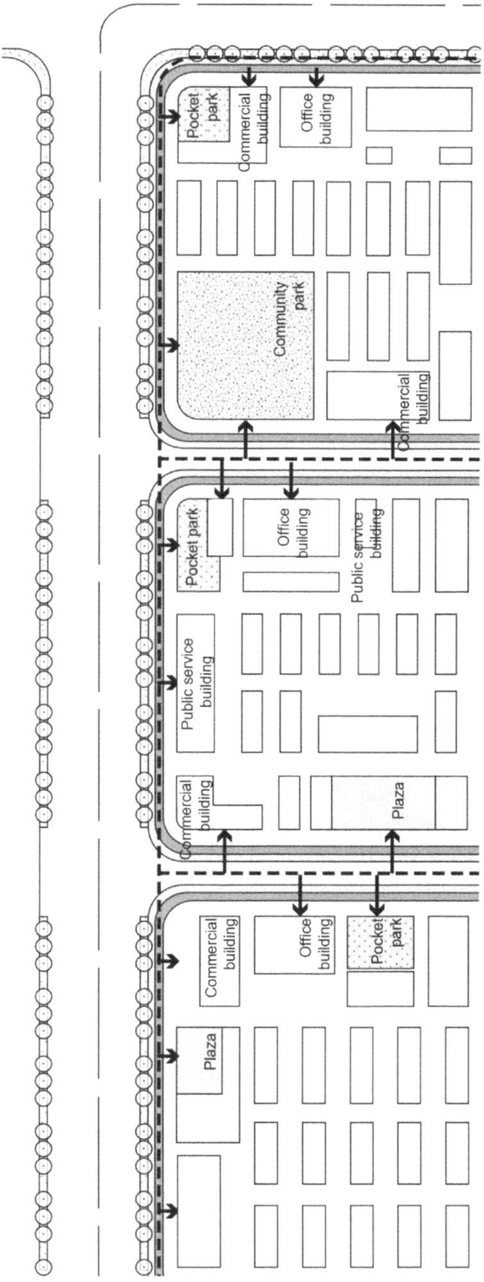

Fig. 3.47 Slow-traffic system connecting pocket parks with surroundings

provide a large number of potential users for parks [40]. For example, in the design parameter system of pocket parks established by Sinou (2013), it not only includes linkage factors such as proximity to surrounding greenways, internal connectivity of the block, and connectivity with other recreational, cultural, and community facilities, but also encompasses accessibility parameters such as accessibility of pedestrian and mobility between the greenways and pocket parks (Table 3.1) [41]. Furthermore, to facilitate the entry of the potential users from the adjacent slow lanes, it is also seemed as an effective strategy to install parking-related facilities, such as bicycle racks, and scenic attractions on park boundary areas.

Additionally, surrounding expressways or trunk roads may negatively affect park visit. Our statistical analysis on the pocket park samples of Chinese cities reveals a notable negative impact of the surrounding trunk road density to pocket park visits at weekend nights [27]. The main negative impact comes from the hindering effect of these roads on the potential park users around, especially diminishing the accessibility of people residing on the opposite side of the road.

3.2.3 Green Spaces

If we examine the relationship between proximately-located green spaces from the perspective of daily recreational service provision, both collaborative and competitive effects may arise among them. The collaborative effect arose from these green spaces commonly attributes to the differentiated service division of them, by which diverse services can be offered to respond to different recreational demands. Consequently, it generates a cumulative benefit that enhances the overall attractiveness and utilization efficiency of the entire green space system. Conversely, the competitive effect is usually produced in the condition that proximately-located green spaces provide similar or homogeneous services, leading to mutual competition in service provision. This competition diminishes the utilization efficiency of each green space and results in resource wastage.

Given the complex interaction mechanism among the proximately-located green spaces, it requires delicate quantitative analysis to identify the interaction mode between green spaces as well as the structure of surrounding demands. Considering the uniqueness of pocket parks, the abundance of nearby green spaces will reduce the scarcity and preciousness of pocket parks in densely populated areas. In particular, if the pocket park is located closely to a large park, its limited size may disadvantage them in competing with the large park in visitor attraction, finally restraining their service performance. This scenario is not conductive to maximizing the overall performance of the park system [42] (Fig. 3.48).

Our studies in Chinese cities also confirmed this fact, which proved that the proximity of the large-sized parks and the greening coverage of surrounding communities (or blocks) show significantly negative effect on pocket park visits (Figs. 3.49, 3.50 and 3.51). However, the impacts of nearby community parks and other pocket parks do not show the equivalent effect, indicating that besides park size and mutual

Table 3.1 Pocket park design parameter system [41]

Main categories	Subcategories	Design parameters
Space	Size	• A small park 50 by 100 feet • Determine the exact size according to anticipated uses
	Space identity	• Determine the function/s of the space • Innovative design concepts • Variety and choice • Greatest play value • Adaptable spaces
	Surfaces	• The walls, those of adjacent buildings, can become vertical lawns • The floor, textural interest and pattern • The ceiling- a canopy of leaves, the solid roof of shade • Avoid blank walls in the space perimeter
	Focal point	• Need of focal point—water features, gazebos, or other structures • Defined edges which may contain a focal point
Environment	Environmental performance	• Maximize the amount of natural shade • Environmentally friendly features: • Pervious surfaces, bio-filter landscaping beds, high-efficiency lighting, solar-powered amenities • Environmental education • Explore aquatic and riparian habitats and stormwater and aesthetic enhancements • Adequate lighting during nighttime
	Location & linkage	• Investigate vacant sites to provide opportunities for temporary public space • Locate the park adjacent to a greenway and the largest concentration of housing • Layout of public open space influences pedestrian flow • Provide paths that lead somewhere • Promote pedestrian use, linkage to network, high-use corner or mid-block link • Linked to other recreational, cultural, and community facilities • Ensure a good flow of park users between the greenway and the park
Society	Users	• Goals: physical fitness, social adjustment, mental and moral improvement, and neighborhood betterment • Play spaces designed for the present population of the area • Comfortable social interaction • Welcoming and appealing design to a diversity of users • Determine major users • Involve the neighborhood in the design process

(continued)

Table 3.1 (continued)

Main categories	Subcategories	Design parameters
	Accessibility	• Easily accessible • Inviting with multiple points of entry • Convenient and safe pedestrian access • Non–traditional locations: rooftops, building facades, or foyers
	Amenities-activities	• Individual sitting replacing bench • Drinking fountains, bicycle racks, trash receptacles, etc • Playground, opportunities for sitting, and open, grassy areas • Range of activities that will activate public space at all times • Maximize the amount of seating • Opportunities for public art
	Safety & maintenix	• Supervision and maintenance • Well buffered from moving cars • Equitable access • Free from barriers to the street • Reduce onerous maintenance requirements

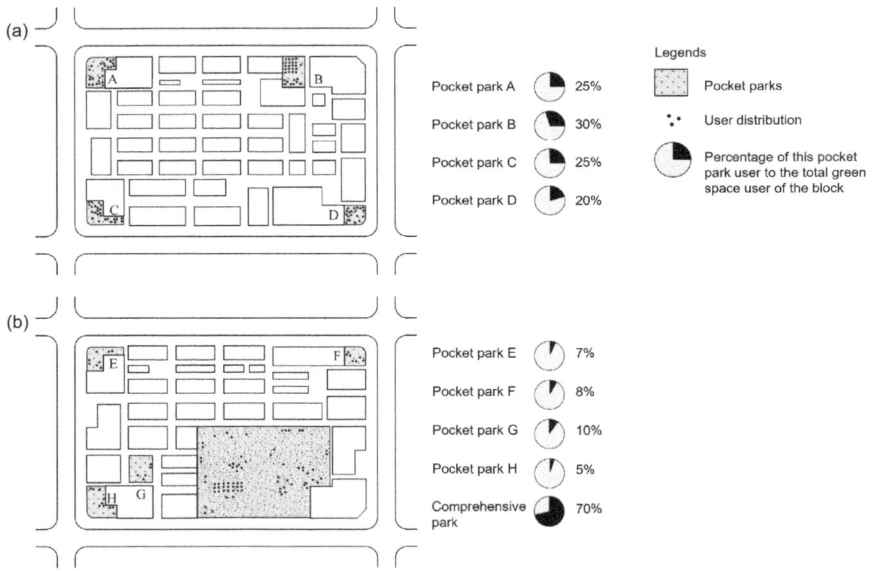

Fig. 3.48 User distribution under different green space interaction mode: (a) User distribution with evenly-distributed pocket parks; (b) User distribution with "a large park + several pocket parks"

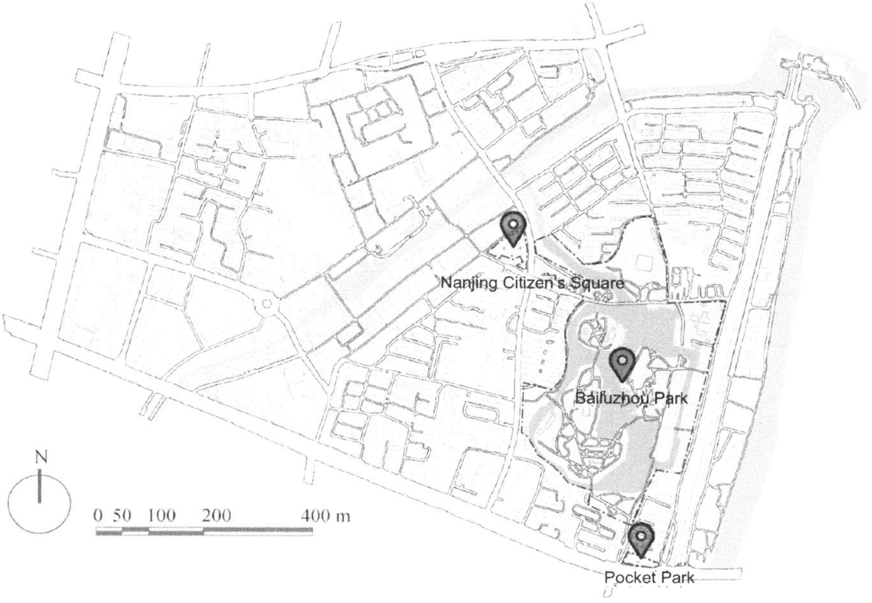

Fig. 3.49 Location of Bailuzhou Park and its surrounding squares and green spaces (Nanjing, China)

distance, there might be more other factors, such as homogeneity and heterogeneity in service mode, facility type, spatial experience, and so on, that influence the interaction among small-sized green spaces [27] (Figs. 3.52, 3.53 and 3.54).

3.2.4 Surrounding Buildings

Commercial, office, cultural, and other public buildings around pocket parks can bring numerous potential users to the nearby parks. Meanwhile, the bright and colorful lighting of these buildings at night can create a pleasant environment for park visit. Moreover, the overall appearance of the surrounding buildings also greatly shapes the experiencing atmosphere in pocket park recreation (Figs. 3.55 and 3.56). Related surveys indicate that the aesthetics of the building appearance, particularly the facade, directly affect the overall perception of park visitors, which further influences the park utilization and vitality [43]. Accordingly, it is not recommended to use the monotonous and boring white walls enclosing the park space. Instead, these building walls can be painted to improve the visual pleasurability and create a more enjoyable experience for park users [14] (Fig. 3.57).

Additionally, if a pocket park is surrounded by high-rise buildings, the orientation and enclosure of the buildings will impact the lighting of the pocket park. Whyte

Fig. 3.50 Activities in Bailuzhou Park (Nanjing, China)

Fig. 3.51 Activities in the pocket park south to Bailuzhou Park (Nanjing, China)

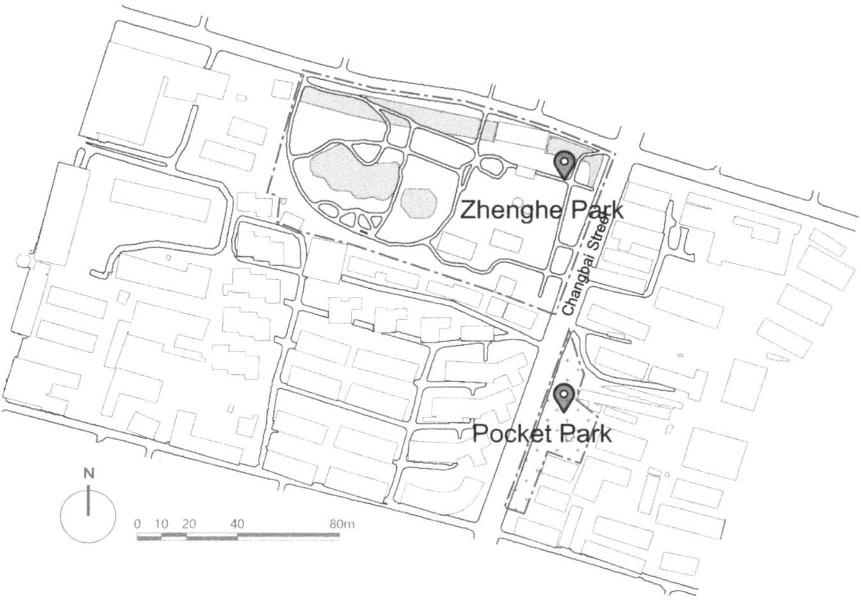

Fig. 3.52 Location of Zhenghe Park and its surrounding pocket parks (Nanjing, China)

Fig. 3.53 Activities in Zhenghe Park (Nanjing, China)

Fig. 3.54 Activities in the pocket park close to Zhenghe Park (Nanjing, China)

Fig. 3.55 Mary 's Square Pocket Park and surrounding buildings (San Francisco, USA) *Source* https://erikwb.net/2013/st-marys-square

Fig. 3.56 Jiangwangmiao Community Pocket Park and its surrounding buildings (Nanjing, China)

Fig. 3.57 Platform Park and its enclosing painted walls (California, U.S.A) *Source* https://www.gooood.cn/platform-park-by-terremoto.htm

Fig. 3.58 Location of crowd activities in Yueshikong Square Pocket Park at 8:00 a.m. (Nanjing, China)

(1980) found that crowds subconsciously followed the move of the area with sunshine in the small public spaces of New York during the cold seasons (Figs. 3.58 and 3.59). However, he also noted that sunshine is not the sole factor that holds the crowds within public spaces. Smart design can also solve this problem in some north-facing public spaces [13]. Nevertheless, pocket parks with enough sunshine can at least provide users with more options. Therefore, pocket parks located on the south or southeast side of buildings usually benefit from better development conditions and natural environments in China (Fig. 3.60).

On the other hand, the degree of enclosure of surrounding buildings also influences the spatial experience within pocket parks. For instance, the space of pocket parks bordered by buildings on their three sides is characterized as low openness and high enclosure, fostering a distinct sense of place which is conducive to the long-time stopover and activities. Those bordered by buildings on parks' two sides strike a balance between space openness and closeness, ensuring a sense of place while connecting with the street space. Pocket parks that are bordered by buildings on their one side or none exhibit significant weakness in place definition. It is usually necessary to further define the space by the means of vegetation to highlight their space boundaries and mitigate the external disturbances, which is helpful to increase the stopovers and stable park uses [44] (Fig. 3.61). Furthermore, our study in Nanjing also showed that pocket parks with clearly-defined boundaries performed better in diversifying the uses of the park space.

Fig. 3.59 Location of crowd activities in Yueshikong Square Pocket Park at10:00 p.m. (Nanjing, China)

3.3 External Socio-Economic Factors

3.3.1 *Potential Users*

Numerous studies have shown that proximity is the most significant factor influencing pocket parks' visits [32, 45]. As surrounding residents are the primary and most stable users of pocket parks, the size and composition of the surrounding residents directly shape parks' utilization pattern. Donahue's study demonstrated that park visitation intensity is significantly associated with the density of the surrounding residents [8]. Our study in Nanjing also proved that the resident size is positively associated with the diversity of pocket park utilization [27].

Furthermore, pocket parks are typically situated within high-density environments, surrounded by numerous existing buildings and facilities apart from apartments and houses, such as offices and commercial facilities, which also provide abundant potential park users. However, these users often visit pocket parks at different time compared to residents. According to our survey, residents tend to utilize pocket parks during early mornings and late afternoons, while officers visit more frequently on weekday mornings, midday, and afternoons and commercial facilities contribute to the park visits on weekends. Moreover, nearby public buildings with main entrances and exits toward to the pocket parks would enhance pocket parks' visit frequency in a more significant way (Fig. 3.62).

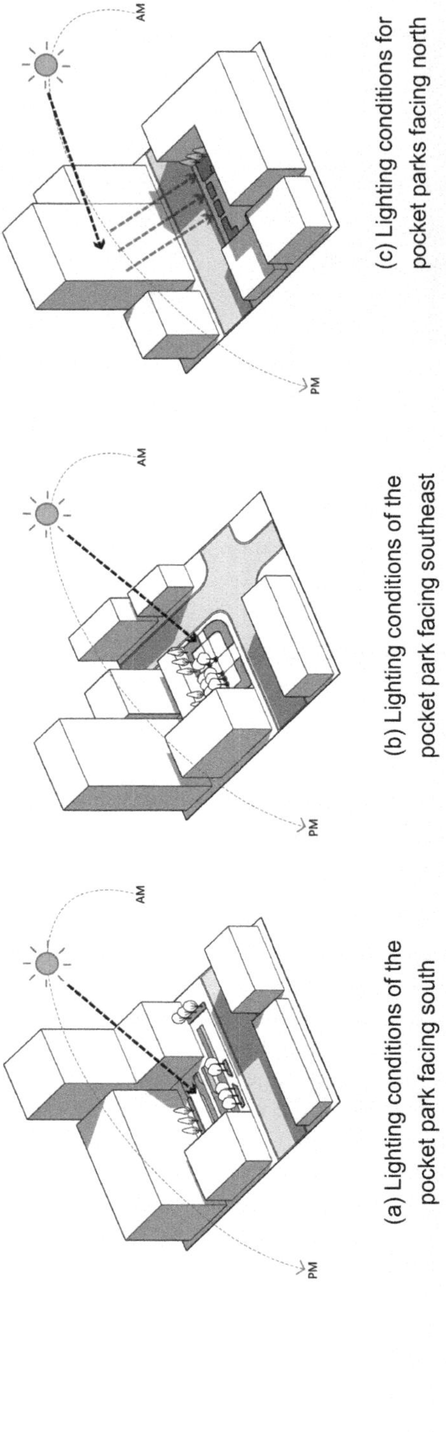

Fig. 3.60 Lighting conditions in different orientation of pocket parks

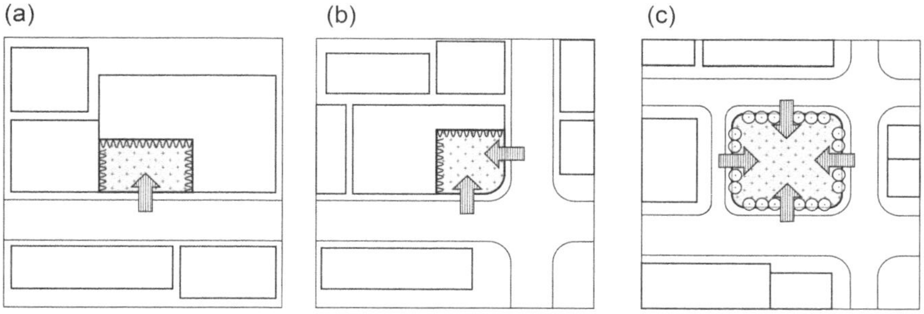

Fig. 3.61 Relationship between building enclosure and spatial experience within pocket parks: (a) Pocket parks bordered by buildings on their three sides; (b) Pocket parks bordered by buildings on their two sides; (c) Pocket parks without surrounding buildings

With the proceeding of refined governance and organic urban renewal, fulfilling the individualized demands of subgroups and individuals received more concerns in green space planning and design. According to our survey on the expected weekly recreation time of four age groups in Yancheng, including the seniors (over 60 years old), middle-agers (>30–60 years old), youths (>10–30 years old), and the children (10 years old and below)], the results showed that the seniors' expected time for weekly recreation was approximately 2.5 times as that of the middle-aged and youth groups, and about two times as that of the children [46] (Fig. 3.63). Thus, if two pocket parks are surrounded by an equal size of potential user group, the user group with a higher proportion of seniors will show a higher demand intensity, probably contributing to a higher frequency of park visits and utilization rate (Fig. 3.64).

Similar differences in park utilization are also led by the differences in groups' social attributes, such as gender, income, religion, ethnicity, and so on. Whyte (1980) observed a heightened critique of women on public space quality [13]. Wendel et al. (2012) found that women exhibit higher sensitivity to lighting, maintenance, and spatial details due to the safety reasons, commonly encountering more restrictions in the visit of pocket parks (Fig. 3.65) [47]. Consequently, the spatial quality and safety measures of a pocket park have a direct impact on its usage rates, particularly if it is frequently visited by females, whereas the significance of this impact may be less pronounced if its main visitors are supposed to be males.

Studies also indicate that income, health status, and language proficiency affect the perceived accessibility of green space. For example, people in high-income neighborhoods tend to access green spaces whose dominant users are their ethnic or cultural peers while those in low-income neighborhoods gravitate towards green spaces where they can find co-enthusiasts with similar activity preferences [48]. Additional studies in English-speaking nations show that individuals from low-income households, poorer health conditions, and non-English-speaking backgrounds perceive limited access to the neighborhood green spaces [3]. Moreover, variations in lifestyle among groups with different faiths and races can influence the demand intensity, visit

Fig. 3.62 Zifeng Building North Entrance Pocket Park (Nanjing, China)

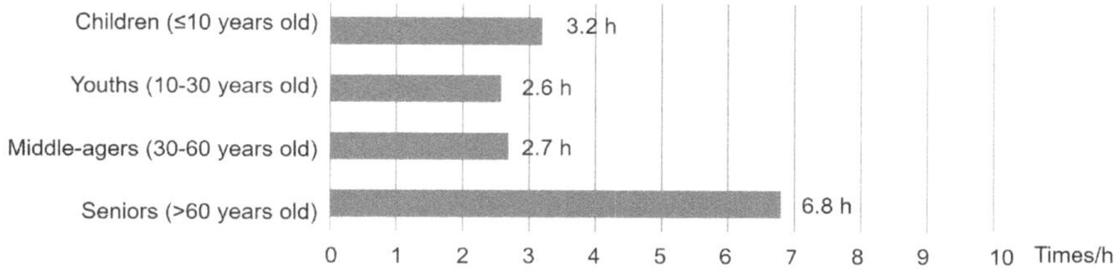

Fig. 3.63 Expected time for weekly recreation of four age subgroups in Yancheng, China

Fig. 3.64 Seniors in Fuqiao East Mini-Park (Nanjing, China)

timing, and utilization pattern of pocket parks, inevitably impacting their service performance.

3.3.2 *Land Value*

Land value exerts an more indirect influence on the pocket park service performance compared to other factors. Boulton et al. (2018) noted that real estate development is tightly related to market demands, thus the development of urban green spaces also necessitates other infrastructure to jointly respond to the demands of existing communities [49]. Consequently, areas with higher land values are typically characterized as higher integrity and quality of infrastructure, for which the installed pocket parks

Fig. 3.65 Men, women, and children in 520 Student Movement Memorial Park (Nanjing, China)

share superior facilities and maintenance standards, demonstrating stronger attractiveness to potential users (Fig. 3.66). In high-density cities, land value also reflects the land scarcity. Our study in Nanjing found that the land value (comprehensive density degree) had an even greater impact on park visits than that of the resident density, further confirming the intrinsic relationship between land value and pocket park service performance [26] (Figs. 3.67 and 3.68).

A similar influence is also evident in the location characteristics of pocket parks, as the location is another key factor in determining land value. According to Zhang & Zhou (2018)'s study in Beijing, green spaces closer to city center experience higher utilization rates [4]. They found that this phenomenon is not only led by the higher population densities of the areas closer to city center, but also attributed to the higher level of investment, delicacy of design, and maintenance, all of which jointly generate the great attractiveness to visitors. Additionally, pocket parks close to the city center have higher proximity to potential users, which is one of the most important determinants on pocket park service performance [27].

3.3.3 Organization and Culture

The core value of pocket parks lies in recreational service. Therefore, it is crucial for pocket parks to effectively meet users' recreational demands to enhance their service performance and satisfaction degree. If the design, construction, and maintenance of

Fig. 3.66 Ginza Sony Park (Tokyo, Japan) *Source* https://www.g- mark.org/award/describe/49496

Fig. 3.67 Yajule Park—pocket park in the areas with relatively high land values (Nanjing, China)

Fig. 3.68 Matai Street Wanrong Park—pocket park in the areas with relatively low land values (Nanjing, China)

pocket parks follows a conventional 'blueprint' mode, it probably leads to an over-rigid service pattern with a lack of adaptability. This will result in a large number of pre-designed spaces and facilities based on a 'design fantasy' that fail to meet the actual demands of the park users and greatly constrain the service efficiency of pocket parks [13]. However, the establishment of an open and dynamic mechanism for design and construction, along with a flexible maintenance and management framework, can provide a systematic guarantee for users by effectively intervening the design, construction, and management of pocket parks. This also creates a good condition for accurately matching the features of the supply side with demand side in public resource allocation and utilization [50].

Additionally, the service status of pocket parks is tightly related to the community culture (i.e. neighborhood relationship, sense of belonging, organization power, and social cohesion et al.). Communities with strong organizational culture often conceive groups or organizations with good cohesion to spontaneously organize public activities within parks, take charge of the park space maintenance and management, and stick up for park's orderly use. In communities with good sense of belonging, pocket parks can serve as a platform for intimate interaction and communication among residents, as well as the main carrier for various group activities, such as square dancing, festival parties, and community gatherings (Fig. 3.69). In contrast, communities that are lack of the belongingness and social cohesion always show a much lower efficiency of pocket park utilization compared to the former ones, even when they possess the higher population density and better facilities (Fig. 3.70).

Fig. 3.69 Yingtuo Community Pocket Park (Nanjing, China)

Fig. 3.70 Wanguofu Community Southeast Mini-Park (Nanjing, China)

References

1. Geng H, Zhang Y, Chi J et al (2024) What affects the satisfaction, preferences, and visitation of pocket parks? Evidence from Shanghai. J Outdoor Recreat Tour 100764. https://doi.org/10.1016/j.jort.2024.100764

2. Akinpar A (2016) Factors influencing the use of urban greenways: a case study of Aydın, Turkey. Urban For & Urban Green 16:123–131. https://doi.org/10.1016/j.ufug.2016.02.004
3. Wang D, Brown G, Liu Y (2015) The physical and non-physical factors that influence perceived access to urban parks. Landsc Urban Plan 133:53–66. https://doi.org/10.1016/j.landurbplan.2014.09.007
4. Zhang S, Zhou W (2018) Recreational visits to urban parks and factors affecting park visits: evidence from geotagged social media data. Landsc Urban Plan 180:27–35. https://doi.org/10.1016/j.landurbplan.2018.08.004
5. Zhang W, Li J (2023) A quasi-experimental analysis on the causal effects of COVID-19 on urban park visits: the role of park features and the surrounding built environment. Urban For & Urban Green 82:127898. https://doi.org/10.1016/j.ufug.2023.127898
6. Xiao Z, Zhang C, Li Y et al (2024) Community park visits determined by the interactions between built environment attributes: an explainable machine learning method. Appl Geogr 172:103423. https://doi.org/10.1016/j.apgeog.2024.103423
7. Chen Y, Liu X, Gao W et al (2018) Emerging social media data on measuring urban park use. Urban For & Urban Green 31:130–141. https://doi.org/10.1016/j.ufug.2018.02.005
8. Donahue ML, Keeler BL, Wood SA et al (2018) Using social media to understand drivers of urban park visitation in the Twin Cities, MN. Landsc Urban Plan 175:1–10. https://doi.org/10.1016/j.landurbplan.2018.02.006
9. Kerishnan PB, Maruthaveeran S (2021) Factors contributing to the usage of pocket parks—a review of the evidence. Urban For & Urban Green 58:126985. https://doi.org/10.1016/J.UFUG.2021.126985
10. Danford RS, Strohbach MW, Warren PS et al (2018) Active Greening or Rewilding the city: how does the intention behind small pockets of urban green affect use? Urban For & Urban Green 29:377–383. https://doi.org/10.1016/j.ufug.2017.11.014
11. Sun X, Wang L, Wang F et al (2020) Behaviors of seniors and impact of spatial form in small-scale public spaces in Chinese old city zones. Cities 107:102894. https://doi.org/10.1016/j.cities.2020.102894
12. Peschardt KK, Stigsdotter UK, Schipperijn J (2016) Identifying features of pocket parks that may be related to health promoting use. Landsc Res 41(1):79–94. https://doi.org/10.1080/01426397.2014.894006
13. Whyte WH, Jr. (1980) The social life of small urban spaces. Conservation Foundation, Washington, DC. https://search.library.wisc.edu/catalog/999547142202121
14. Marcus CC, Francis C (1990) People places: design guidelines for urban open space. Van Nostrand Reinhold, New York, NY. https://search.library.wisc.edu/catalog/999624093802121
15. Gehl J (2001) Life between buildings: using public space. The Danish Architectural Press, Copenhagen
16. Luximon Y, Kwong HY, Tai YY (2015) User preferences of urban park seating pattern in Hong Kong. Procedia Manuf 3:4273–4278. https://doi.org/10.1016/j.promfg.2015.07.415
17. Zhou C, Wu Y, Hu Y, et al. (2017) Green spectrum along the city wall--graphic analysis of characteristics and service performance of the Nanjing Ming Dynasty City Wall greenway, 1st edn. Southeast University Press
18. Salih S, Ismail SB, Mseer A (2020) Pocket parks for promoting social interaction among residents of Baghdad City. Archnet-IJAR: Int J Arch Res. https://api.semanticscholar.org/CorpusID:225000166
19. Bernstein FA (2017) New York's Paley Park, which turns 50 this month, is a masterpiece—proof that even a tiny public space can make a difference in a crowded city. Landsc Arch 107(5):122–129
20. Thompson JW (1997) The rebirth of New York City's Bryant Park. Spacemaker Press, Washington, DC. https://search.library.wisc.edu/catalog/999833027102121
21. Gehl J (1996) Public spaces, public life. Danish Architectural Press: Royal Danish Academy of Fine Arts, School of Architecture, Copenhagen. https://search.library.wisc.edu/catalog/9910335760602121

22. Grabiec AM, Łacka A, Wiza W (2022) Material, functional, and aesthetic solutions for urban furniture in public spaces. Sustainability 14(23):16211. https://doi.org/10.3390/su142316211
23. Nordh H, Alalouch C, Hartig T (2011) Assessing restorative components of small urban parks using conjoint methodology. Urban For & Urban Green 10(2):95–103. https://doi.org/10.1016/j.ufug.2010.12.003
24. Ikin K, Beaty RM, Lindenmayer DB et al (2013) Pocket parks in a compact city: how do birds respond to increasing residential density? Landsc Ecol 28(1):45–56. https://doi.org/10.1007/s10980-012-9811-7
25. Ishibashi S, Akasaka M, Koyanagi TF et al (2020) Recognition of local flora and fauna by urban park users: who notices which species? Urban For & Urban Green 56:126867. https://doi.org/10.1016/j.ufug.2020.126867
26. Zhou C, An Y, Zhao J et al (2022) How do mini-parks serve in groups? a visit analysis of mini-park groups in the neighbourhoods of Nanjing. Advance online publication, Cities. https://doi.org/10.1016/j.cities.2022.103804
27. Zhou C, Fu L, Xue Y et al (2022) Using multi-source data to understand the factors affecting mini-park visitation in Yancheng. Environ Plan B: Urban Anal City Sci 49(2):754–770. https://doi.org/10.1177/23998083211029660
28. Jacobs J (2011) The death and life of great American cities (50th anniversary edition). Modern Library, New York. https://search.library.wisc.edu/catalog/9910656313202121
29. American Society of Landscape Architects (2012) Lafayette Greens: urban agriculture, urban fabric, urban sustainability. Retrieved April 22, 2024, from https://www.asla.org/2012awards/073.html
30. Ministry of Housing and Urban-Rural Development, PRC (2016) Code for the Design of Public Parks (GB 51192–2016)
31. Holy-Hasted W, Burchell B (2022) Does public space have to be green to improve well-being? an analysis of public space across Greater London and its association to subjective well-being. Cities 125:103569. https://doi.org/10.1016/j.cities.2022.103569
32. Hamstead ZA, Fisher D, Ilieva RT et al (2018) Geolocated social media as a rapid indicator of park visitation and equitable park access. Comput, Environ Urban Syst 72:38–50. https://doi.org/10.1016/j.compenvurbsys.2018.01.007
33. Zhou C, Xie M, Zhao J et al (2022) What affects the use flexibility of pocket parks? evidence from Nanjing. Land. Advance online publication, China. https://doi.org/10.3390/land11091419
34. Development Bureau of the Government of the Hong Kong (2011) Design and management guidelines for public open space in private developments. Hong Kong
35. San Francisco Planning Department (2010) Walking, bicycling & public space on market street. San Francisco Planning Department, San Francisco
36. Aziz N (2016) Potentials of creating pocket parks in high density residential neighborhoods: the case of Rod El Farag, Cairo city. https://api.semanticscholar.org/CorpusID:197675203
37. Aziz NAAE (2017) Pocket Park Design in Informal Settlements in Cairo City, Egypt. Landsc Arch RegNal Plan 2(2):51–60. https://doi.org/10.11648/j.larp.20170202.12
38. Harper J (2009) Planning for recreation and parks facilities: predesign process, principles, and strategies. Venture Publishing, Inc., State College, Pa. https://search.library.wisc.edu/catalog/9910467099002121
39. Gibson H, Canfield J (2016) Pocket parks as community building blocks: a focus on Stapleton. CO. Community Dev 47(5):732–745. https://doi.org/10.1080/15575330.2016.1220965
40. Chang PJ (2020) Effects of the built and social features of urban greenways on the outdoor activity of older adults. Landsc Urban Plan 204:103929. https://doi.org/10.1016/j.landurbplan.2020.103929
41. Sinou M (2013) Parameters contributing to the design of a successful urban pocket park.
42. Harnik P (2010) Urban green: innovative parks for resurgent cities. Island Press, Washington, DC. https://search.library.wisc.edu/catalog/9910092652002121
43. Nordh H, Hartig T, Hagerhall CM et al (2009) Components of small urban parks that predict the possibility for restoration. Urban For & Urban Green 8(4):225–235. https://doi.org/10.1016/j.ufug.2009.06.003

44. Ashihara Y (1970) Exterior design in architecture. Van Nostrand Reinhold, New York. https://search.library.wisc.edu/catalog/999598194802121
45. Rigolon A (2017) Parks and young people: an environmental justice study of park proximity, acreage, and quality in Denver, Colorado. Landsc Urban Plan 165:73–83. https://doi.org/10.1016/j.landurbplan.2017.05.007
46. Zhou C, Zhang Y, Fu L et al (2021) Assessing mini-park installation priority for regreening planning in densely populated cities. Sustain Cities Soc 67:102716. https://doi.org/10.1016/j.scs.2021.102716
47. Wendel HEW, Zarger RK, Mihelcic JR (2012) Accessibility and usability: green space preferences, perceptions, and barriers in a rapidly urbanizing city in Latin America. Landsc Urban Plan 107(3):272–282. https://doi.org/10.1016/j.landurbplan.2012.06.003
48. Łaszkiewicz E, Kronenberg J, Marcińczak S (2021) Microscale socioeconomic inequalities in green space availability in relation to residential segregation: the case study of Lodz. Poland. Cities 111:103085. https://doi.org/10.1016/j.cities.2020.103085
49. Boulton C, Dedekorkut-Howes A, Byrne J (2018) Factors shaping urban greenspace provision: a systematic review of the literature. Landsc Urban Plan 178:82–101. https://doi.org/10.1016/j.landurbplan.2018.05.029
50. The Trust for Public Land (2020) Pocket Park Toolkit. The Trust for Public Land, San Francisco. Retrieved April 15, 2024, from https://www.tpl.org/resource/pocket-park-toolkit

Chapter 4
Park Group

In high-density environments, the concentration of recreation demands and scarcity of recreation resources present an irreconcilable contradiction [1]. Although pocket parks can provide new recreation opportunities, the limited individual capacity constrains their capability to effectively alleviate the serious shortage of outdoor recreational spaces in high-density areas, especially using the mode of randomly installation [2]. Correspondingly, integrating the dispersedly-distributed pocket parks into a serving group to jointly respond to the concentrated recreational demands of residents offers an effective strategy [3].

The construction of pocket park groups serves dual purposes: first, it fosters a cumulative effect by aggregating small green spaces, thereby increasing the green space size and recreational opportunities; second, it lays the groundwork for establishing a sequence or system of recreational services and makes up for the capacity limitations of single pocket park by activating the complementary effect between proximately-located pocket parks and enhancing the overall effectiveness of green space system [4].

According to China's *Standard for Urban Residential Area Planning & Design (GB 50180-2018)*, in the redevelopment of areas with scarce land resource and high density which show difficulties in meeting the minimum size threshold of an individual residential park, a "multi-points distributional mode" is recommended to substitute one large-sized residential park with several distributed small-sized green spaces [5]. *The 15-min Community Life Circle Planning Guidelines of Shanghai* also proposes to create a public space network through a systematic installation of the green spaces at community-level and below [6]. These regulations and guidelines have created favorable conditions for establishing pocket park groups within Chinese cities.

C. Zhou, *Principles and Methods of Pocket Park Planning and Design*, Urban Sustainability, https://doi.org/10.1007/978-981-96-5801-5_4

4.1 Goals

4.1.1 Equity in Resource Allocation

As the concept of 'Environmental Justice' is being popularized in current urban development and renewal [7], the spatial distribution of urban green spaces, as a key approach to allocate public resource, are increasingly being recognized as a matter of equity and justice. For instance, Dai (2011) [8] and Boone et al. (2009) [9] discovered that African-American in US cities showed less accesses to green spaces compared to the whites; Wüstemann et al. (2017) [10] and Kabisch and Haase (2014) [11] found significant differences in the access to green space among groups with different social attributes (e.g. age, income, family composition, etc.) in European cities; Richards et al. (2017) [12] and Tan and Samsudin (2017) [13] observed significantly lower per capita green space area in high-density and lower-income districts of Southeast Asian cities.

In China, due to the continuous increase of population density led by decades of urbanization while the neglect of the synchronous development of green spaces in the early urban construction, it can be widely observed a serious imbalanced distribution of green spaces between the newly developed areas and the old city or downtown areas in Chinese cities. The development of pocket park groups provides precious chances for reallocating public resources, thereby mitigating the disparities of the green space distribution among the different areas within the cities (Figs. 4.1 and 4.2).

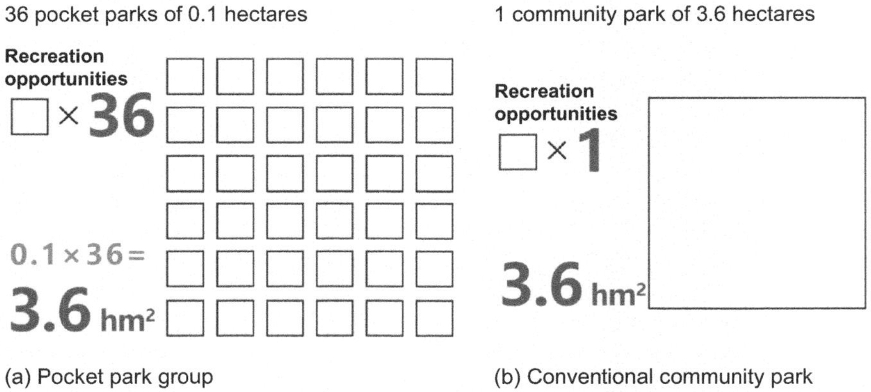

Fig. 4.1 Comparison of size and recreational opportunities between pocket park groups and community parks

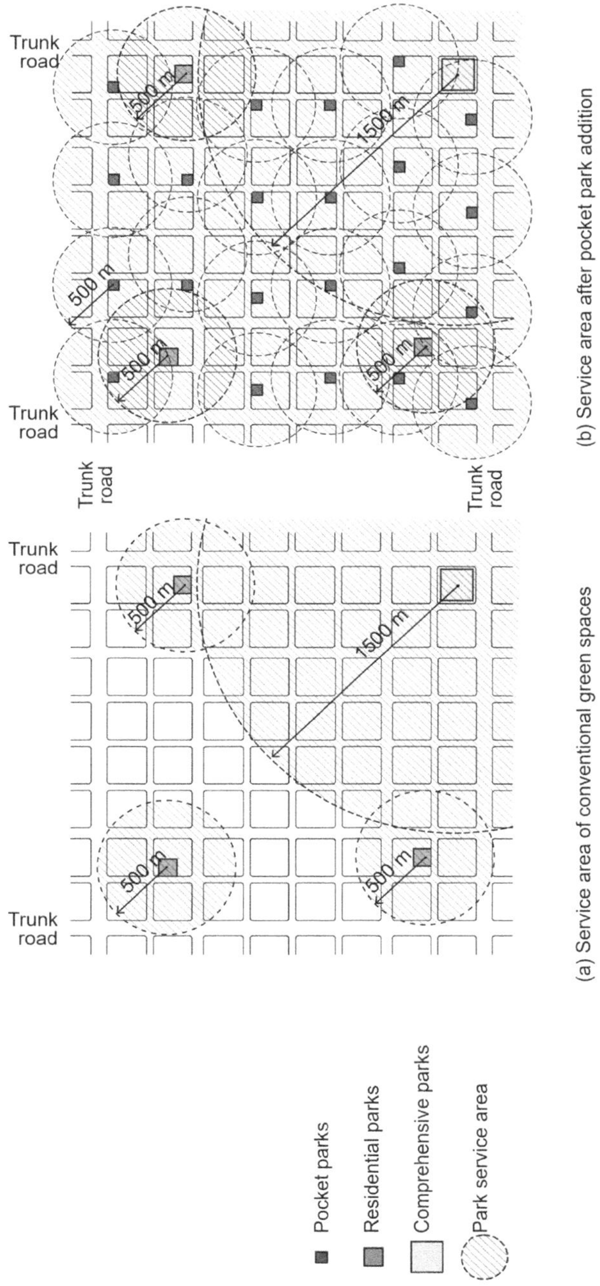

Fig. 4.2 Complementary effect of pocket park groups to conventional park service

4.1.2 Response to Personalized Demand

Nowadays, the theory of "New Public Service" is dominating the urban public management, which defines the role of government as a servant and brings the city management and planning into the stage of fine governance [14]. This requires a much higher respect to the individualized demands and preferences of different social groups in planning, compared to the traditional mode of "New Public Management" which defines the role of government as a steersman and focuses more on the quantity and integrity of public product supply [15].

Accordingly, traditional urban green space planning often followed the "supply-oriented" mode, which easily neglected the requirement differentiation on demand side, leading to mismatches between resource supply and demand. In response, pocket park groups, characterized by decentralized distribution and proximate service provision, actually offer an innovative solution framework to shift the green space planning from the "supply-oriented" paradigm into the "demand-oriented" paradigm [16]. Pocket parks' proximity in service provision enables a miniaturization and specialization of the target service groups, creating a good condition for parks to provide customized and precise service to meet the specific demands of the subdivision groups (Fig. 4.3). The essence of this initiative is to respond to the social granularity refinement of service targets by decomposing the spatial granularity of service units. Consequently, a pocket park group can provide diverse recreational services by differentiating parks' dominant function and facility types within the group.

4.1.3 Construction of Daily Recreational System

The traditional park service model is mostly shaped based on the multi-ring demand system, which is characterized by a clear hierarchical division of the service targets. The ideal model includes: (1) country parks and forest parks at the urban periphery, that mainly serve for regional recreational activities in short- or long-term vacations; (2) urban parks (or comprehensive parks), which primarily serve for citywide recreational activities in weekends or short-term vacations; (3) neighborhood and pocket parks, categorized as community recreational spaces, which mainly serve for people's daily recreational activities [17] (Fig. 4.4).

Based on this ideal model, parks of different classes are designated varied service radii, as evidenced by the green space classification system and planning guidelines of UK [18], Japan [19], and China [20] (Table 4.1). The primary purpose of establishing this hierarchical system is to facilitate a balanced spatial distribution of parks, ensuring that the service scope of parks of each class can fully cover the target area. However, the uneven distribution of residents across the city spaces makes it challenging to seamlessly match with the demand distribution through an evenly distributional mode of parks. This challenge is particularly evident in high-density

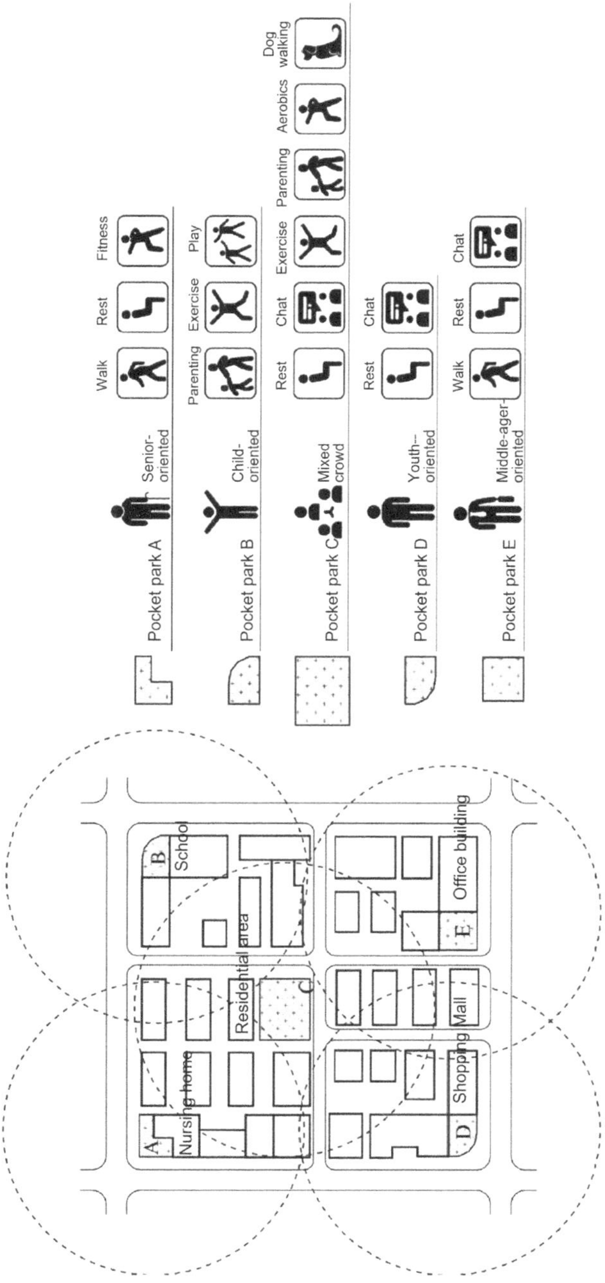

Fig. 4.3 Customized services of pocket parks responding to the specific demands of different groups

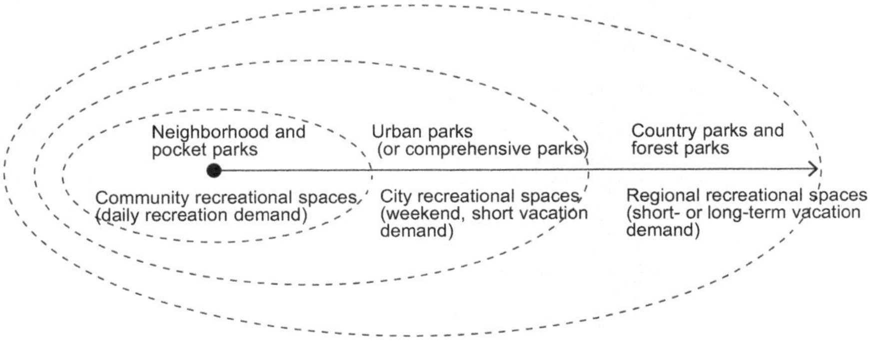

Fig. 4.4 Traditional park service model

areas, where the disparities of supply and demand of outdoor recreational spaces are highly significant, resulting in severe lack of park provision locally and dysfunction of the classic hierarchical model. Given the scarcity of outdoor recreational spaces in densely populated areas, the urban parks and part of the specialized parks (like relic parks and theme parks), that are originally designated to serve for the demands of holiday or weekends, have to prioritize fulfilling the daily recreational demands of nearby residents in their service process.

Therefore, we can see a prevailing stream occurring in many high-density areas, which is characterized by substituting the classic hierarchical model with the flatten model in green space service. One prominent evidence is that the large-sized parks are spontaneously or purposely divided into multiple small recreational service units in daily use to meet the daily recreational demands of surrounding residents, especially in the areas where small-sized parks at community class are lacked or absent. As observed by Whyte (1980) in his study in New York, "*Central Park is also a host of small spaces, and people experience it as such.*" [21] Our research in Nanjing also proved that the playgrounds and other activity spaces distributed around the central lake of Xuanwu Lake Park, the most famous urban park of the city, can be decomposed into multiple recreational service units on weekdays, which can separately serve for the nearby residents, just as community-class parks do (Fig. 4.5).

Moreover, multiple small green spaces can be integrated into an organic service sequence or system to diversify the park services, thereby compensating for the absence of large-sized parks (Fig. 4.6). For instance, Zion (1969) proposed utilizing the interstices between buildings and street corner spaces to create a pocket park system in response to the shortcomings of the traditional park hierarchy model found in the downtown area of New York City [22]. This flatten service model aims to maximize the service potential and efficiency of each green space in the high-density environment. However, it also raises higher requirements on green space planning and design, particularly emphasizing the necessity of constructing pocket park groups.

Table 4.1 Green space classification system and planning guidelines of UK [18], Japan [19], and China [20]

Classification	Public open space classification of London			Urban park classification of Japan			Public park classification of China		
	Class	Size (hm^2)	Service radii (km)	Class	Size (hm^2)	Service radii (km)	Class	Size (hm^2)	Service radii (km)
Above community levels	Regional parks	400	3.2–8	Landscape parks	>10	60 min to reach	Comprehensive parks	≥50	>3
	Municipal parks	60	>3.2	Sports parks	>10	30 min to reach		20–50	2–3
	District parks	20	1.2	General parks	>10	2.0		10–20	1.2–2
Community and lower levels	Local parks	2	0.4	Neighborhood parks	>5	1.0	Residential parks	5–10	0.8–1
	Small open spaces	<2	<0.4	Juvenile parks	>0.8	0.6		1–5	0.5
	Pocket gardens	<0.4	<0.4	Junior parks	>0.5	0.5	Mini parks	0.2–1	0.3
				Toddler parkss	>0.2	0.25			

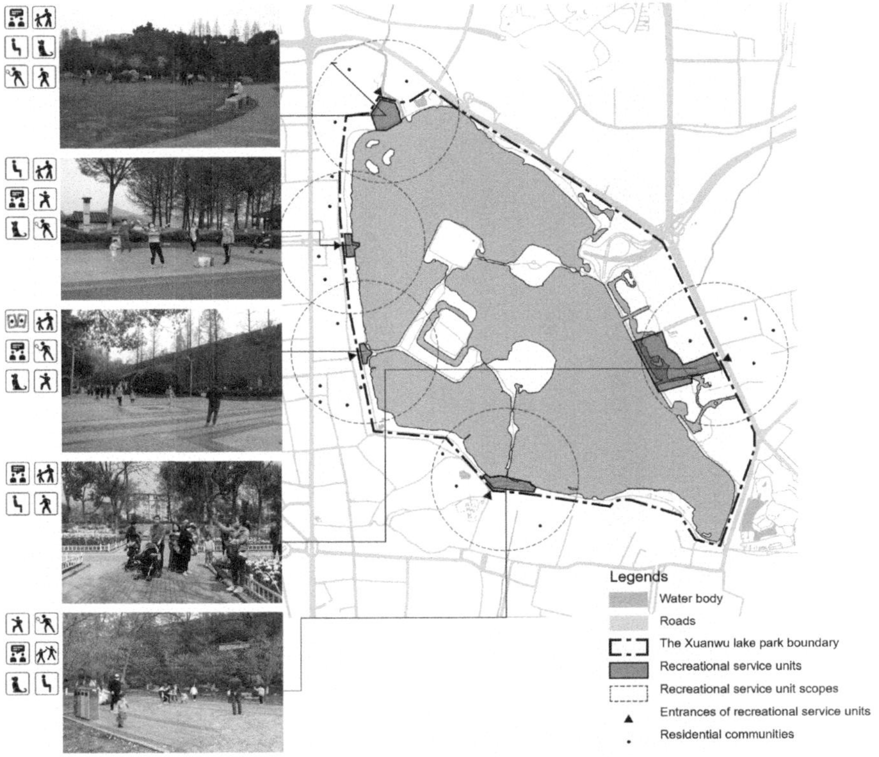

Fig. 4.5 Multiple recreational service units within Xuanwu Lake Park that serve for nearby residents in daily recreation (Nanjing, China)

4.2 Definition

4.2.1 Functional Definition

The professional term "group" typically refers to the collection of people or things that are located close together or are classed together. In urban planning theory, the concept of "group" first emerged in the theory of regional development which focuses more on the geographical attributes. For instance, "city group" denotes cohesive collections of cities in a specific geographical scope, consisting of the cities with different functions, sizes, and classes [23]. Although cities within the group demonstrate spatial and administrative independence, they are tightly interconnected through socio-economic relationships. The study of city groups, drawing from the theories of human ecology, central place, location, urban geography, and general system, mainly focuses on investigating the relative relations on function [24], size [25], and spatial distribution [26] among cities.

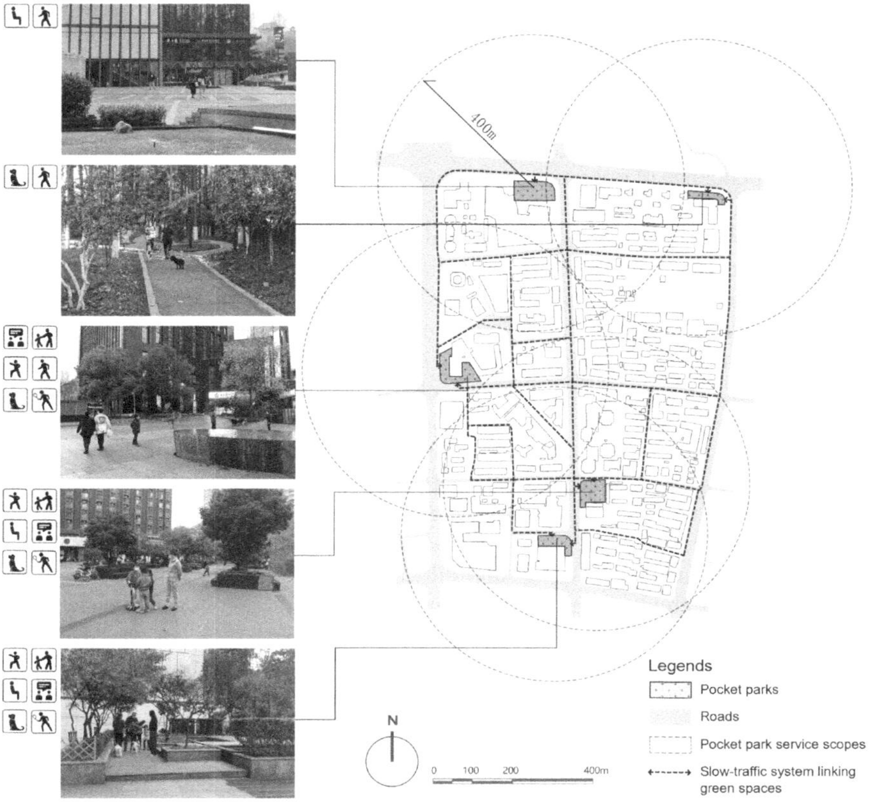

Fig. 4.6 Recreational service sequence integrating multiple pocket parks within a block (Nanjing, China)

Although the interconnections among pocket parks are less intricate than those among cities, the definition and analytical approach of city groups can provide a useful tool for the analysis of pocket park groups. According to the essence of the group, pocket parks that can be categorized into the same group should own three commonalities, including the homology of service objects, the independence of operation and management, and the relevance of function.

The primary service objects of pocket parks are local residents. Once the resident density of an area is so high that the recreational demand is too intense to be fulfill by a single pocket park, constructing a pocket park group can deconcentrate the centralized demand into multiple distributed parks. This is the primary purpose of establishing pocket park groups. Accordingly, although the service scope of pocket parks in a group is different in space, their main service objects should be homologous, such as the residents within the same spatial unit (e.g. block, neighborhood, and so on). For instance, although China's *Standard for Urban Residential Area Planning and design (GB 50180-2018)* [5] recommends installing multiple small-sized green spaces to compensate for the absence of the larger community parks, these small-sized green spaces should share the same target service area (community or neighborhood).

The independence of operation and management denotes that pocket parks within a group should function as distinct service units, particularly in terms of operation and management in order to avoid mutual interference. Specifically, each pocket park should provide recreational services independently without being constrained by others in the group. For instance, they should have separate entrances and exits and can be operated autonomously. Operational independence is also the prerequisite for the service diversification and customization of the pocket parks within a group.

The relevance of function entails that pocket parks within a group should mutually support and complement in the daily service. For instance, each park has a unique focus in terms of service positioning, spatial arrangement, and facility configuration, which jointly construct an integrated service sequence or system. Functional relevance of pocket parks is difficult to form through an isolate design of individual pocket park which defines the park's function separately and randomly. Instead, early interventions on function division and allocation are necessary during the group construction to establish the mutual relevance. This functional relevance is also vital for maintaining a high level of overall service effectiveness of the groups.

4.2.2 Spatial Definition

The service scope of a single pocket park is typically defined to match the "5-Minute Life Circle" (with a radius of 200–400 m). To satisfy the homology requirement of service objects, the distance between the two farthest pocket parks within the group should be controlled within the distance of two preset service radius ($\leqq$ 800 m) (Fig. 4.7). If the mutual distance of the pocket parks is longer than this, they probably serve for the residents from different areas, thereby showing weak interconnection in between.

According to the service mechanism of pocket parks and relevant practical experience, in order to spatially define the pocket park groups, the references that can be commonly taken are the community boundaries and slow traffic barriers. The community boundaries serve as management limits. The term "community" originated as a sociological concept, referring to "*a human group in a definite territorial area, whose members have common activities and experiences*" [27]. With the development of functional zoning and settlements in modern cities, "community" evolved into a planning concept with defined size and scale characteristics [28]. In Chinese cities, communities are the subordinate units of subdistricts ("JIEDAO") and are managed by grassroots self-governing organizations entitled as "resident committees", which have their own management boundaries in space. For instance, Shanghai's *15-min Community Life Circle Planning Guidelines* [29] states that the

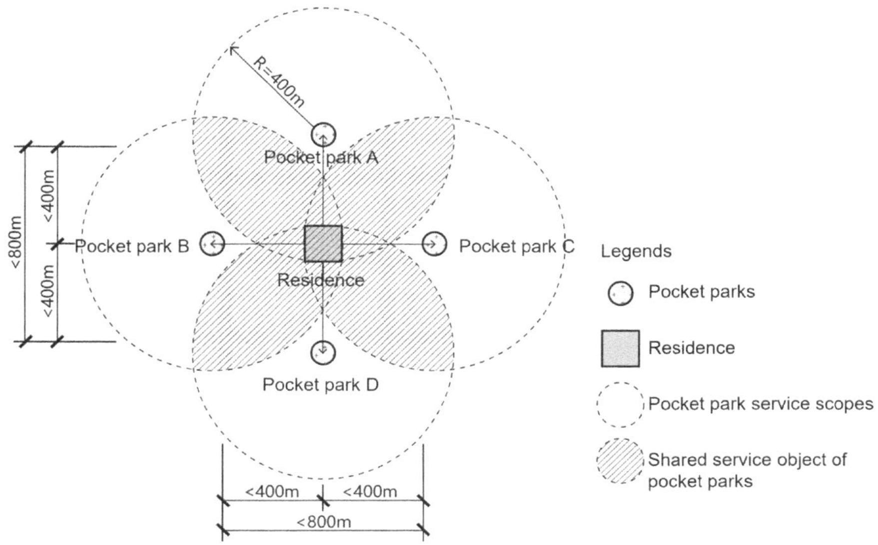

Fig. 4.7 Layout of Pocket parks sharing the homology requirement of service objects

primary service objects of the public space network in the community, with pocket parks as the mainstay, are the residents of the community (Fig. 4.8).

Using community boundaries to define the service scope of the pocket park groups shows two benefits: first, it fully consists with the statistical units of demographic data (e.g., population number, population structure, and households), facilitating an efficient classification and analysis of the recreational demands of groups' service units; second, it can easily integrates the pocket park group into the community's entire public facility system in the planning process, facilitating an overall balance, implementation, management, and post evaluation and monitoring in the planning and utilization. However, as a type of administrative limits, they cannot guarantee a continuous and pedestrian-friendly environment within these boundaries, raising a risk of realistic spatial division of park visitation within the community led by the transportation cutoffs. Furthermore, residents' access and use of pocket park groups basically follow the objective laws of environmental perception and behavior rather than these imperceptible management boundaries. Both of these may lead to the disparities between the area enclosed by community boundaries and the real service scope of the pocket park groups.

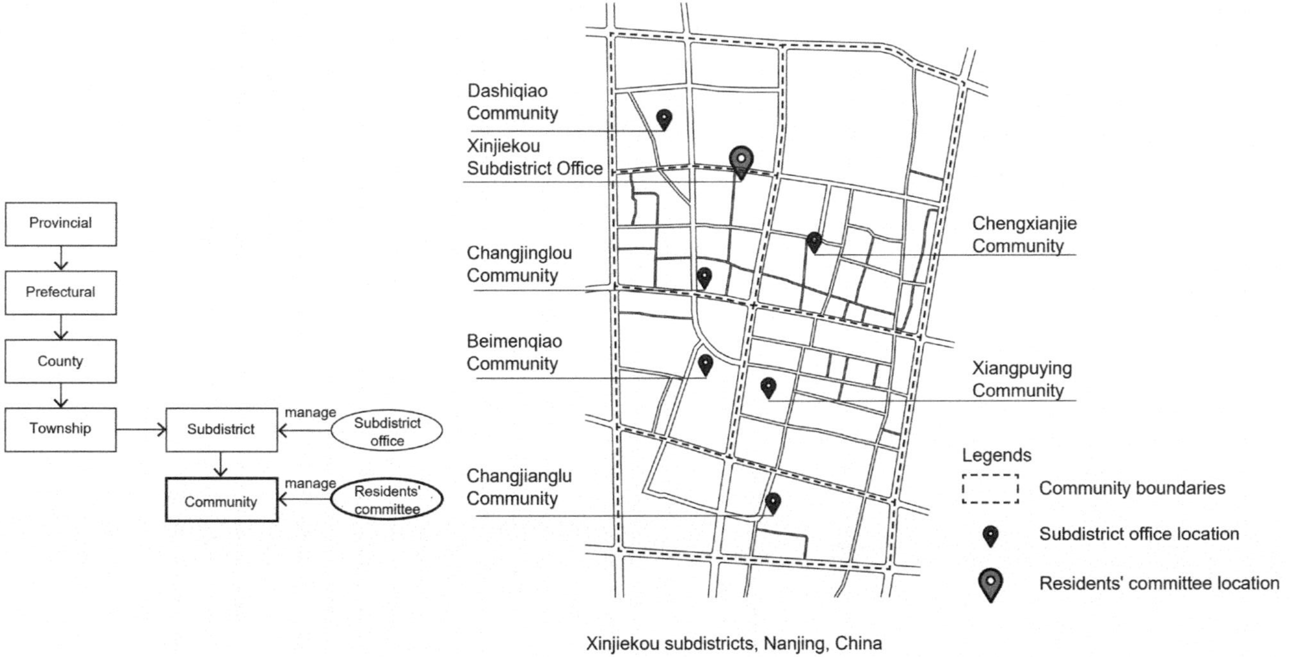

Fig. 4.8 "Community" and Chinese administrative system (Xinjiekou Subdistrict, Nanjing, China)

The cutoffs of slow traffic (such as walking and cycling) provide another spatial reference for the division of the group service scope. According to the China's *Standard for Urban Residential Area Planning and Design (GB 50180-2018)* [5], the boundaries of 15-min, 10-min, and 5-min pedestrian-scale neighborhoods employ the urban roads of different classes to define, respectively, through which the scale of the neighborhood classes are in accordance with the block size enclosed by the corresponding class roads (Fig. 4.9). To guarantee the functional relevance of pocket park groups, it is crucial to connect parks through a continuous slow traffic network. As urban trunk roads are proved to significantly hinder the slow traffic and park visits, especially for pedestrians, it is advisable to avoid the cutoffs led by trunk roads in the establishment of pocket park groups, which raises high requirements on the pedestrian-friendly environment. In most high-density areas, the interval distance of trunk roads are usually less than 800 m, aligning with the requirement of the farthest distance of mutual pocket parks within a group to maintain the homology of the group's service object. Accordingly, using slow traffic cutoffs to define the service scope of pocket park groups demonstrates considerable feasibility and effectiveness, fitting well with the real service scope of the pock park groups. However, in planning practice, the boundaries of units defined by this mode might not conform to the community boundaries. Consequently, their statistics need to be remade while their implementation and post monitoring require more cross-community coordination, greatly increasing the work difficulty and complexity.

As the advantages of either one of the aforementioned methods showed potentials to compensate for the disadvantages of the other, it can reach a more balanced and reasonable result to integrally apply them in the planning practice. One recommended strategy is to initially delineate the group service scope using the community boundaries, and then adjust the boundaries according to the morphology of the trunk road network (Fig. 4.10).

In summary, the ideal service unit of pocket park groups should exhibit two essential characteristics: first, to guarantee the homology of group's service objects, the side length of the unit should not exceed 800 m, which equals to an area that should be smaller than 70 hectares and akin to the size of a typical community in densely populated cities of China; second, to maintain a pedestrian-friendly environment in which is free of slow traffic cutoffs like trunk roads, it limits the unit size for group serving between the sizes of 5-min and 10-min pedestrian-scale neighborhoods, that are outlined in the China's *Standard for Urban Residential Area Planning and Design (GB50180-2018)* [5] (Fig. 4.11). These criteria are also resemblance to the "Neighborhood Unit" model proposed by Clarence Perry in the 1920s, which typically encompass an area of approximately 160 acres (65 hectares), with trunk roads serving as boundaries to prevent the traversal of abundant vehicles in order to maintain the pedestrian-friendly environment within the unit [30, 31] (Table 4.2).

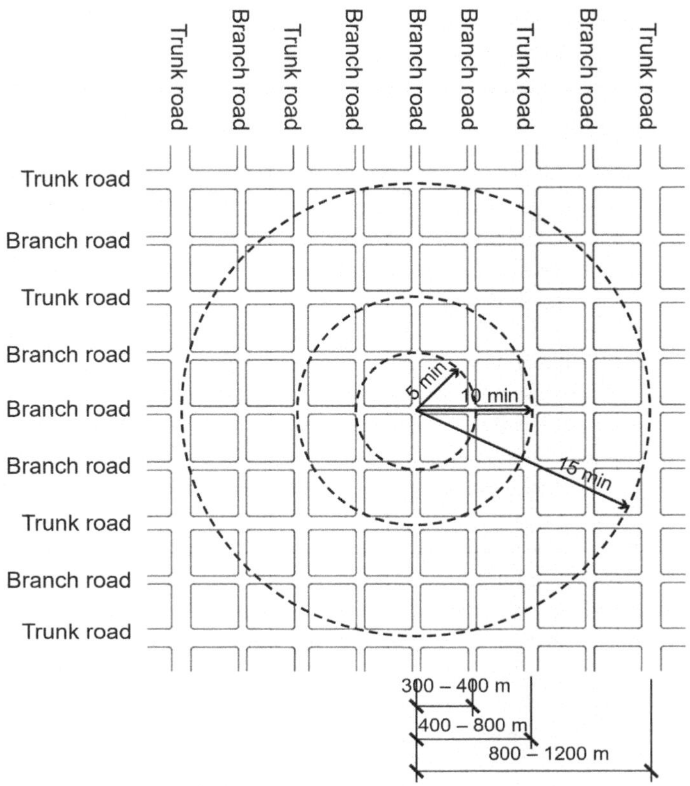

Fig. 4.9 Pedestrian-scale neighborhood system

4.3 Structure

4.3.1 Service Structure

(1) Service Division Modes

The functional interrelationship of pocket parks within a group greatly determines their service structure. In practice, according to the primary user groups or dominant usage patterns, service of parks in a group can be divided by three modes: horizontal, vertical, and composite.

In the horizontal division mode, the service of each park caters to the demand of one distinct primary user group or predominant usage pattern, resulting in the significant differentiation of functional definition of the parks. This mode is highly applicable for the service units that accommodate diverse social groups while each group concentrates in different areas, such as a unit with both neighborhoods of youth dormitories and senior apartments. This division mode enables a clear identification

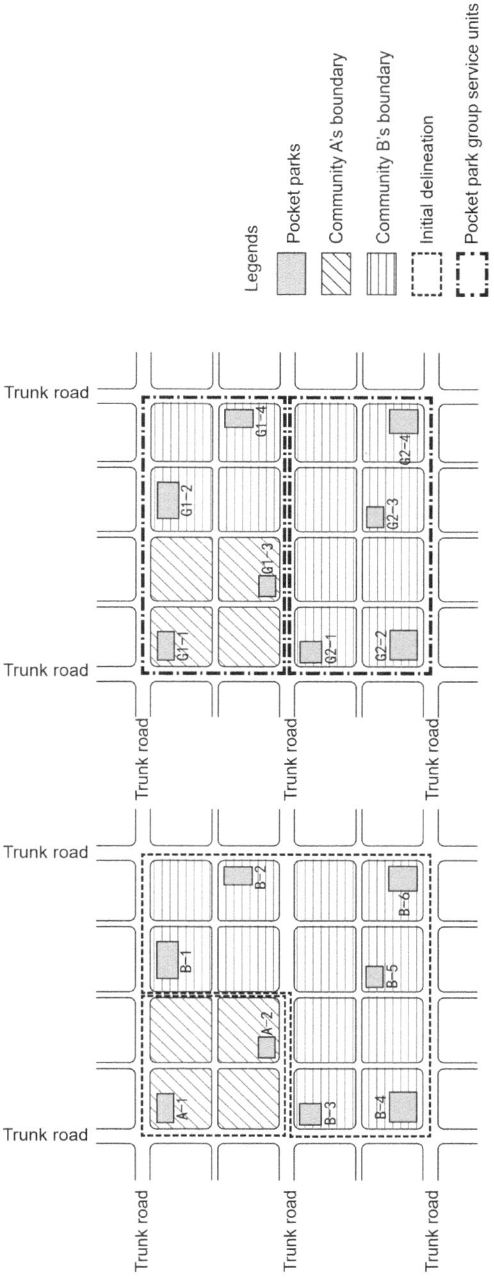

(a) Initial delineation based on community boundaries

(b) Boundary adjustment according to the morphology of trunk road network

Fig. 4.10 The method of spatial definition of the pocket park groups

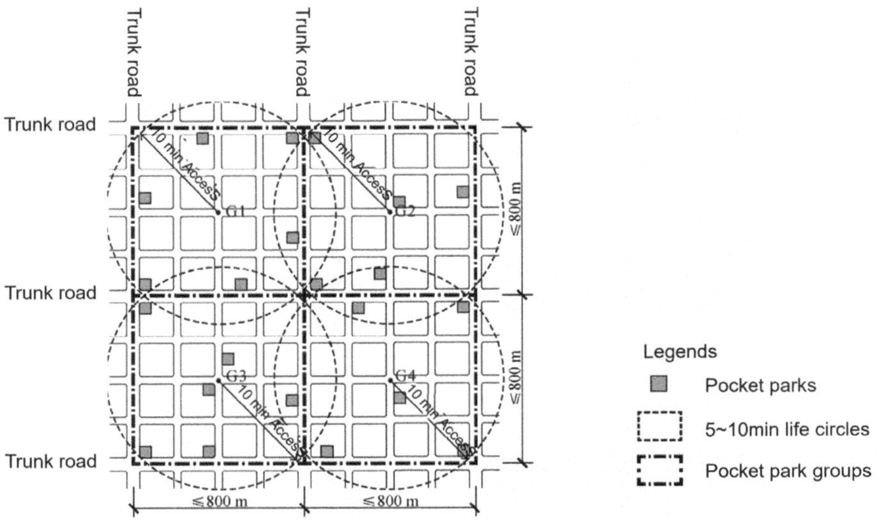

Fig. 4.11 Service units of pocket park groups

Table 4.2 Neighbourhood units and their characteristics

Service unit		Boundaries	Site size (hm^2)	Population (people)	Walking distance (m)
Pedestrian-Scale Neighborhoods of Standard for Urban Residential Area Planning and Design (GB 50180-2018)	15-min pedestrian-scale neighborhoods	Trunk roads or site boundaries	200–400	50,000–100,000	800–1200
	10-min pedestrian-scale neighborhoods	Trunk roads, branch roads, or site boundaries	40–80	15,000–25,000	400–800
	5-min pedestrian-scale neighborhood s	Roads of branch and above grade or site boundaries	10–30	5000–12,000	300–400
Clarence Perry's "Neighborhood Unit"		Trunk roads	About 65	3000–4000	<800

of the primary service targets and usage patterns of each pocket park according to the uniqueness of the nearby demands (Fig. 4.12).

In the vertical division mode, pocket park groups entail distinct hierarchical characteristics. Pocket parks, that serve as the group center in the high-level, typically offer centralized recreational services with large capacities. These parks can

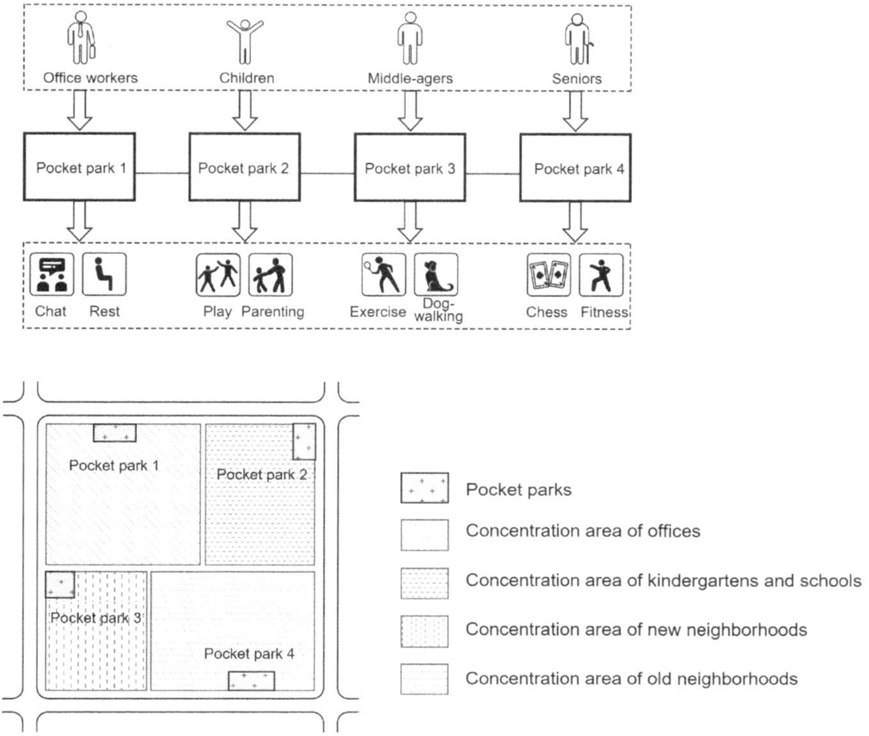

Fig. 4.12 The horizontal division mode of park service

commonly support group activities and allow for diverse usage patterns, thus demonstrating high inclusiveness. Pocket parks at the lower levels usually serve as specialized nodes that offer specified recreational service, which are tailored for the demand of the specific group or usage pattern. This mode is highly applicable in the units where different social groups merge together in space and the development conditions of pocket parks differ significantly across different areas (Fig. 4.13).

In the composite division mode, either the primary user group or predominant usage pattern is not preset for each park within the group. Instead, through a meticulous analysis of the composition, using time, and usage patterns of various potential user groups, it adopts strategies such as time-sharing, primary-secondary compatibility, and flexible utilization to construct a composite service system. This system aims to maximumly activate the service potential of each pocket park within the group, jointly providing systematic recreational services for various user groups at the same or different periods. This mode is suitable for the service units with concentrated recreational demands and limited space for pocket park development (Fig. 4.14).

(2) Types and Characteristics

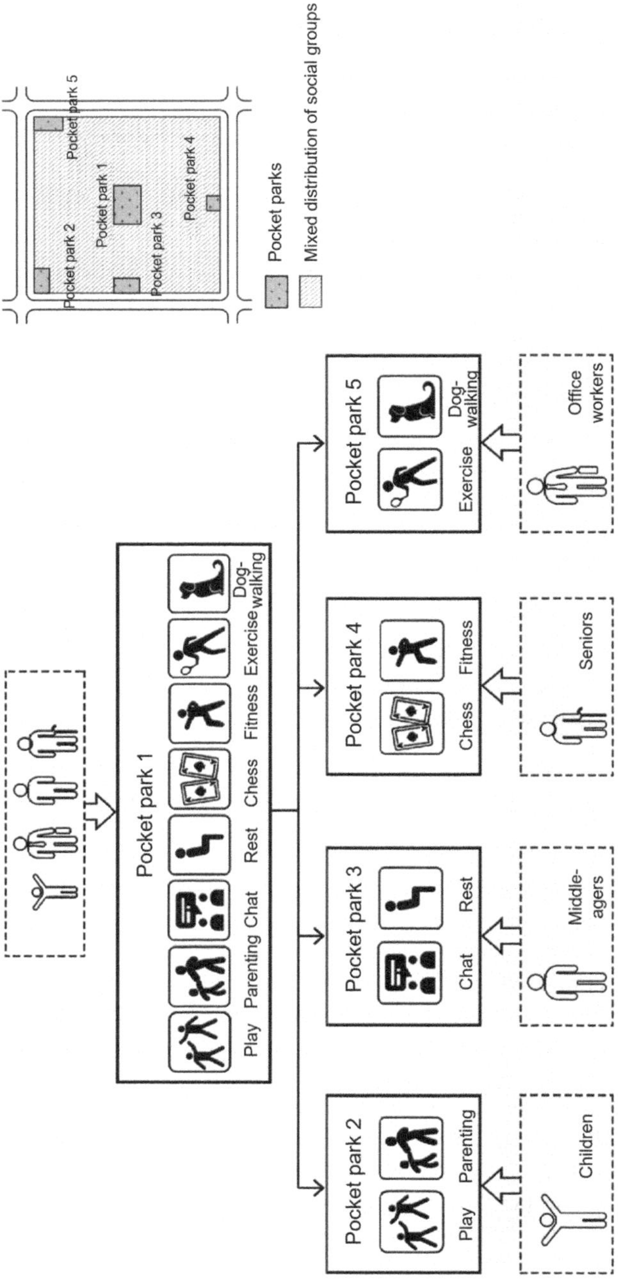

Fig. 4.13 The vertical division mode of park service

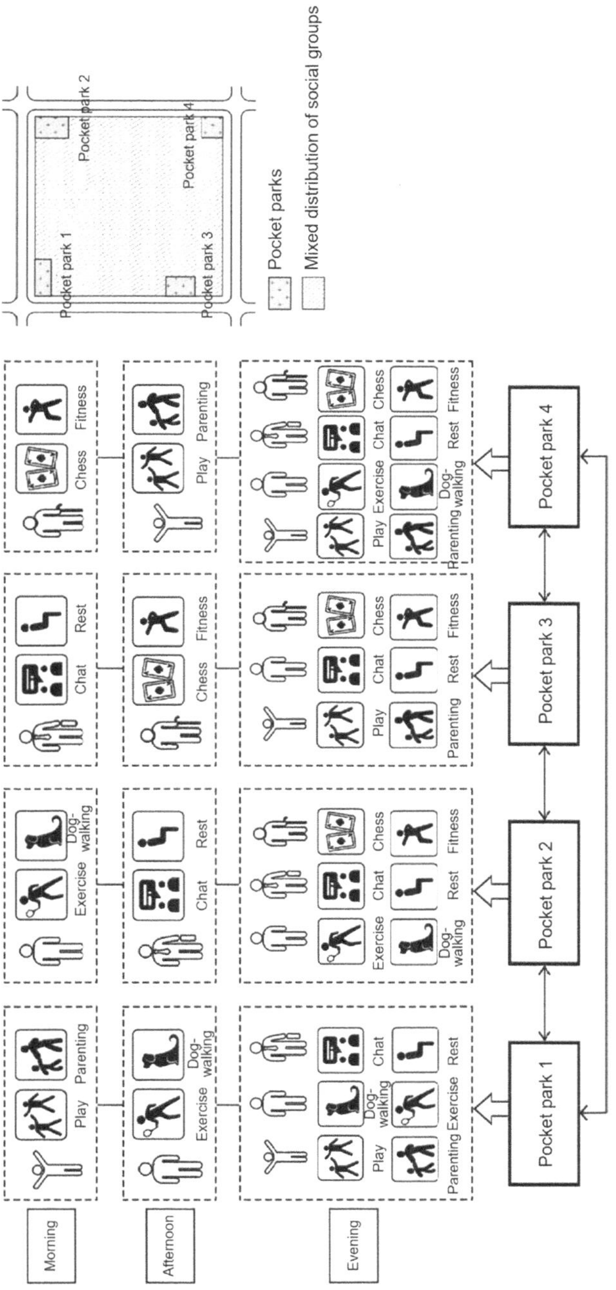

Fig. 4.14 The composite division mode of park service

According to the relative service effectiveness of pocket parks within a group, the service structure of pocket park groups can be classified into the types of single-center, multi-centers, and no-center (Table 4.3). Given the tiny size of all pocket parks, the size cannot be employed as an absolute reference for defining the "service center" in a park group. Instead, the service effectiveness, as commonly indicated by the number of park visitation (park visits), serves as the key indicator. Accordingly, apart from the factors on size and capacity, a pocket park designated as "service center" often shows superior location to be conveniently connected and accessed and more inclusiveness to meet the activity demands of diverse user groups.

In the single-center service structure, the "center" pocket park undertakes the main service task within the park group, whose park visits typically accounts for more than 40% of the group's total visits. This type of service structure may be observed in either vertical or composite division modes (Fig. 4.15). In this structure, the center park not only generates greater recreational attractiveness, but also shows better connectivity with the different areas in the service unit. In some occasions,

Table 4.3 Types and characteristics of service structures

Type	Service division modes	Core characteristics	Advantages	Disadvantages
Single-center service structure	The vertical and composite division modes	• The center park's visits account for more than 40% of the group's total visits	• Promotion of concentrated resident engagement in diverse recreational activities	• Imbalances service status between pocket parks, • Capacity overload and service quality decline at center park
Multi-center service structure	The vertical and composite division modes	• Center parks' visits achieve at least 70% of the group's total visits; • Center parks are fewer than non-center parks	• Function and service division among center parks	• Imbalanced service effectiveness of parks • Insufficient utilization of non-center parks
No-center service structure	The horizontal and composite division modes	• No individual park's visits exceed 40% of the group's total visits	• Balanced service effectiveness of parks • Strong fluidity and mutual support	• Excessively even distribution of recreational spaces among parks that might be short of the sites of collective activities for the dominant user groups

the over-aggregation of the potential users around the center park also leads to the formation of this structure. In this structure, non-center parks commonly serve as the complements to the center park, functioning as the service nodes in the system.

The multi-center service structure typically comprises two or more service centers while the number of center parks should be less than the number of non-center parks. The park visits of the center parks should be at least 70% of the group's overall visits. This structure is commonly seen in the vertical or composite division modes (Fig. 4.16). The main drivers of the formation of this structure type are the concentration of the potential users in multiple locations. Sometimes, this structure proves significant differentiation in size, function, and amenity configurations of the pocket parks within a group, leading to a severe imbalance in service capacity and quality.

The no-center service structure usually has no obvious service centers, in which the visits of each park usually does not exceed 40% of the group's overall visits. In this structure, the service effectiveness is relatively balanced among parks and the potential users are usually evenly distributed within the service unit. In most scenarios, the dominant user group or usage pattern of each pocket park within the group is clearly defined and there is a high degree of mutual support between parks in the system. Therefore, this structure is commonly found in the horizontal and composite division modes (Fig. 4.17).

It is worth noting that the service structure of pocket park groups may also change dynamically over time due to the instability and variation of green space service status. For instance, a group with multi-center service structure according to its total service data of a whole day may show a single-center or no-center structure in some specific hours (Fig. 4.18). Conversely, a group proving no-center service structure according to the service data of accumulated period may display evident service centers in several short moments (Fig. 4.19).

(3) Measurement

The measurement of the service structure of pocket park groups can be seen as an approach to quantitatively describe the contribution of individual parks to the overall service effectiveness of the group and the overall characteristics of the group structure. Each park's contribution to group's service effectiveness can be assessed through the ratio of park visits, while the overall characteristics of the group structure can be gauged using metrics such as the Entropy Index (EI) [32] or Herfindahl–Hirschman Index (HHI Index) [33] (Table 4.4). These two metrics, originally derived from the measurements of the market share concentration in economics, were used to signify the market concentration levels [34]. A higher index value indicates greater market concentration, with the value of 1 representing a complete monopoly [35]. This principle can also be applied in assessing the service structure of pocket park groups. Specifically, a higher value reflects the concentration of park visits toward the center parks. The value of 1 indicates an absolute monopoly of park visits of one single park, while the value of 0 denotes a complete even distribution of park visits across all parks, signifying an ideal centerless structure. Although calculated

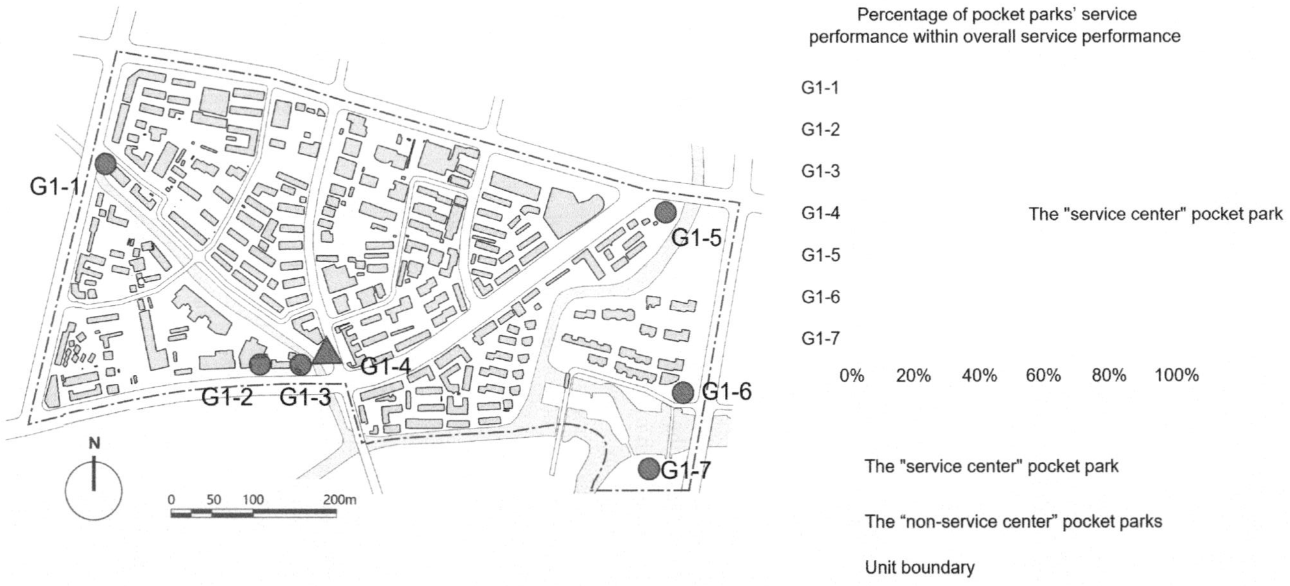

Fig. 4.15 Example of the single-center service structure: Baixia Road Taiping South Road Southeast Block (Nanjing, China)

Fig. 4.16 Example of the multi-center service structure: Central Road East Beijing Road Northeast Block (Nanjing, China)

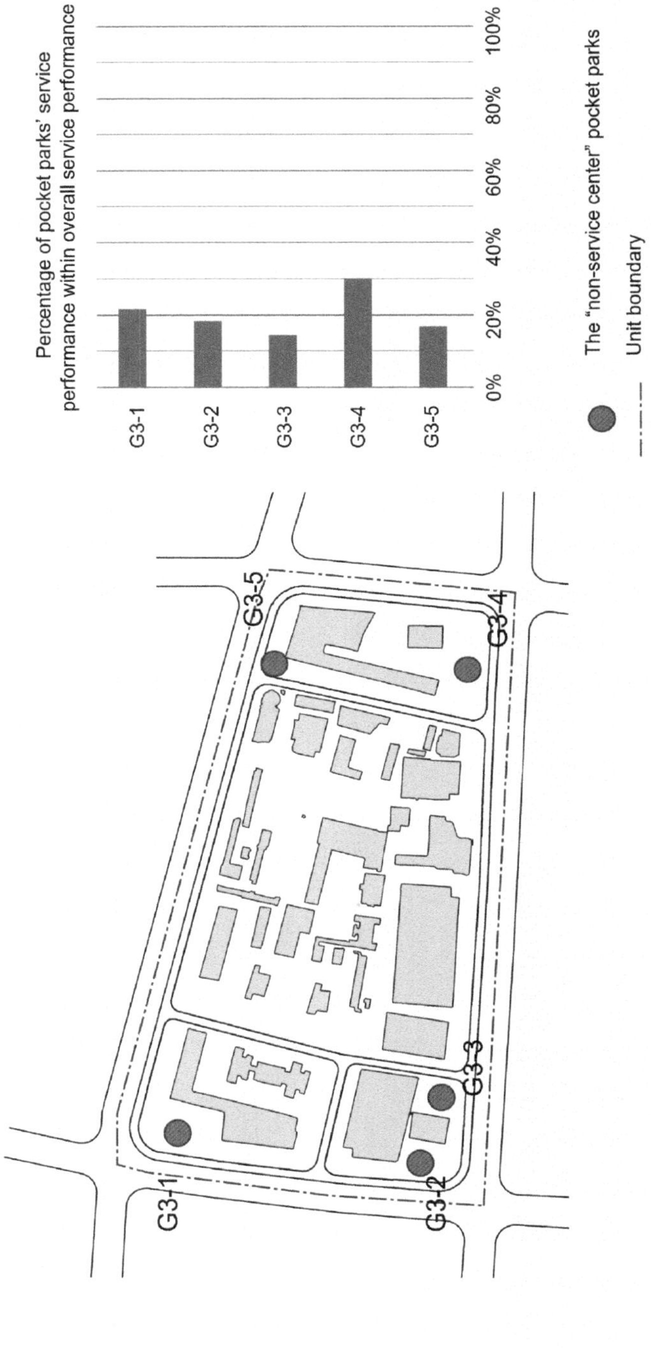

Fig. 4.17 Example of the no-center service structure: Hongwu North Road Changjiang Road Southeast Block (Nanjing, China)

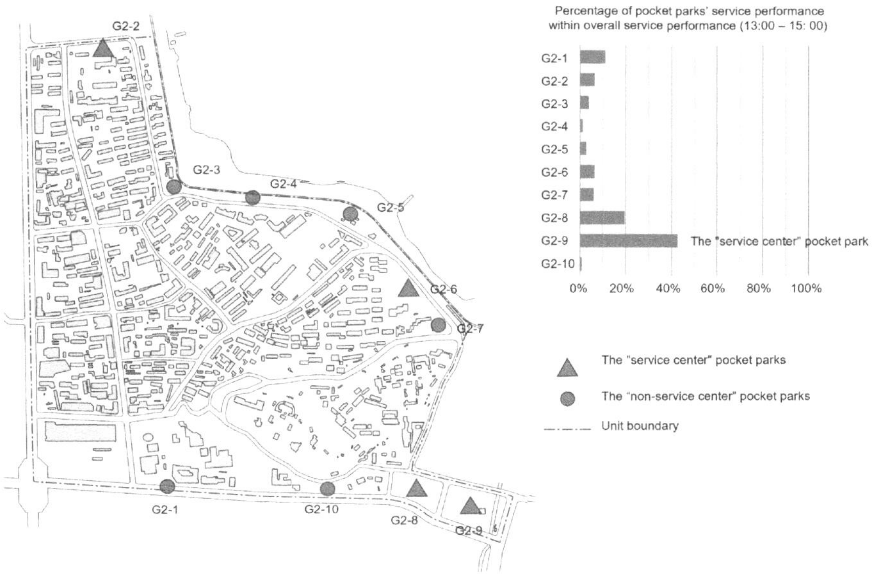

Fig. 4.18 Example of pocket park group with multi-center service structure showing single-center service structure in specific period: Central Road East Beijing Road Northeast Block (Nanjing, China)

in different ways, the EI and HHI indicators serve similar roles and may be selected interchangeably in practice (Fig. 4.20).

4.3.2 Spatial Structure

(1) Types and Characteristics

The structural constituents of pocket park groups in space mainly encompass pocket parks and the links between them. Pocket parks mainly function as the service centers or nodes within a group, shaping the fundamental framework in space. Parks' mutual links primarily entail the slow-traffic lines, determining both the flow of the visitors and the spatial relationships among the parks. The spatial structure of pocket park groups can be categorized into four types based on the distribution of parks and their linkage patterns, including the types of ring, radial, linear, and network (Table 4.5).

For park groups in ring structure, pocket parks are typically positioned at the periphery of the service unit and linked by a ring shaped slow-traffic road. This structure is commonly observed in the access-limited blocks or neighborhoods whose building density is high in the center area and open spaces concentrated at the

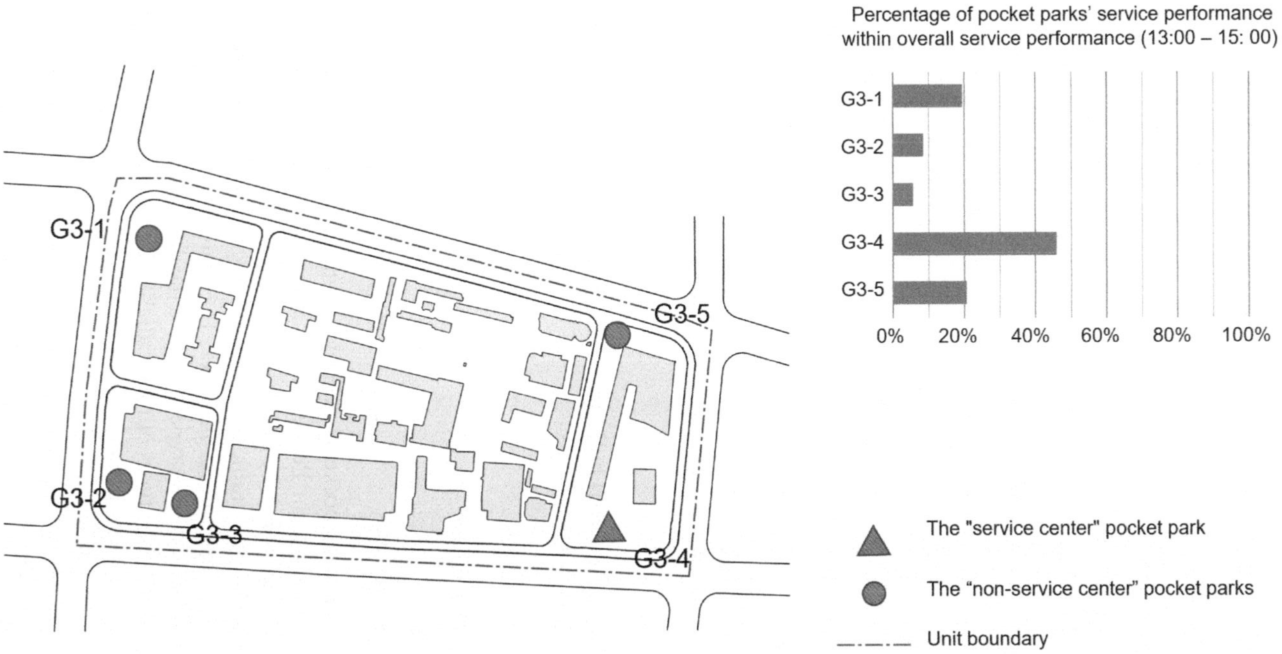

Fig. 4.19 Example of pocket park group with no-center service structure showing single-center service structure in specific period: Hongwu North Road Changjiang Road Southeast Block (Nanjing, China)

Table 4.4 Measurement metrics of service structure

Item	Metrics	Formula	Description
Overall balance	EI	$\mathrm{EI} = -\left[\sum_1^k P_i \mathrm{Ln}(P_i)\right]/\mathrm{Ln}(k)$	The higher EI value, the more even distribution of park visits across all parks
		EI is the entropy index of the group, k is the number of pocket parks in the group, P_i is the proportion of visitors of the pocket park *i* relative to the total visitor number of the group	
	HHI	$\mathrm{HHI} = \sum_1^k (N_i/N)^2$	The value closer to 0, the more even distribution of park visits across all parks
		HHI is the Herfindahl–Hirschman Index, k is the number of pocket parks in the group, N_i is the visitors of the pocket park *i*, *N* is the total visitor number of the group	
Each park's contribution	Ratio of park visits	$C_i = N_i/N$	The higher value, the more important the pocket park is in the group
		C_i is the contribution of pocket park *i*, N_i is the visitors of the pocket park *i*, *N* is the total visitor number of the group	

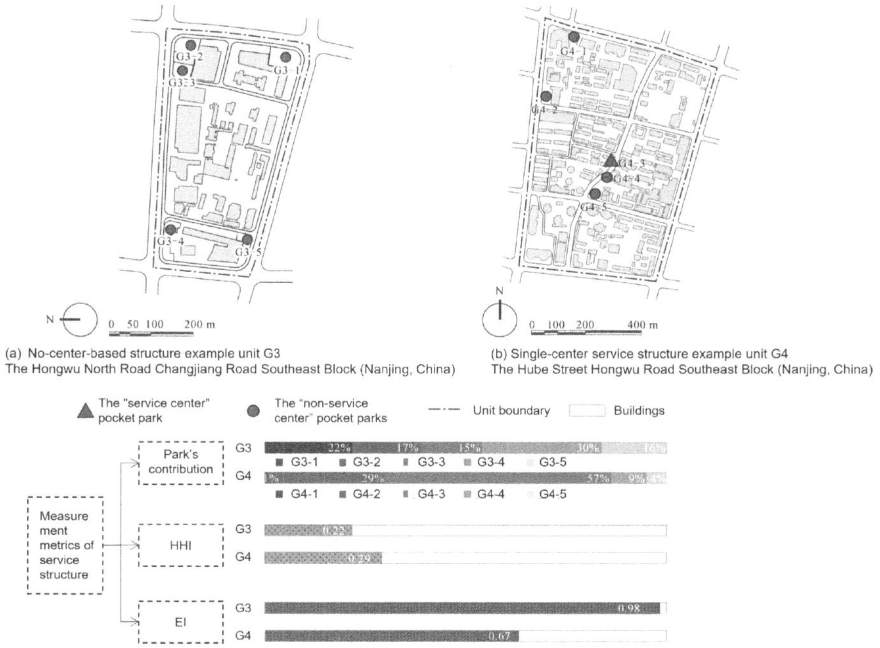

Fig. 4.20 Application of service structure metrics of pocket park groups

Table 4.5 Types and characteristics of spatial structures

Type	Diagram	Applicable environment	Advantages	Disadvantages
Ring structure	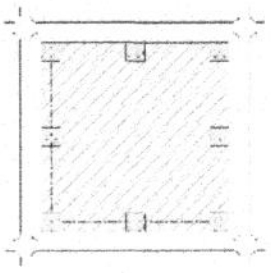	• Access-limited block or neighborhoods with high building density in the center area and open spaces at the boundary areas	• A continuous and enclosed service sequence, facilitating activities such as daily walking, running, cycling, dog walking, and so on	• Relatively singular links between parks and the simplicity of their interaction patterns
Radial structure	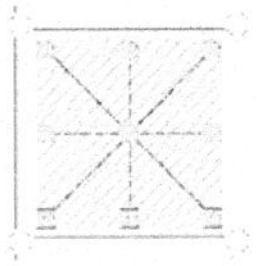	• Neighborhoods with central open spaces	• Extension of the service of pocket park groups to the internal area of the unit • High green space accessibility	• Challenge in forming coherent and closed park service sequence • Great discrepancies in service efficiency between central and peripheral pocket parks
Linear structure	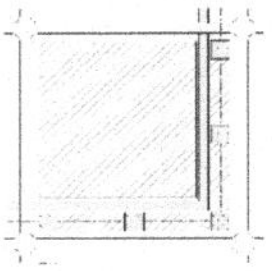	• Neighborhoods with little internal open space	• A continuous and non-closed service sequence, reserving extension interfaces to connect with the green space system of the adjacent neighborhoods or blocks	• Significant discrepancies in park accessibility of the residents in different areas within the unit • Relatively limited connection patterns between the pocket parks
Network structure	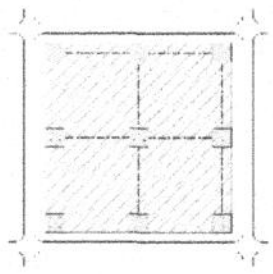	• Open neighborhoods with more internal open space and rich slow-traffic network	• Diverse service sequences and circulation patterns • High accessibility	• Rigorous conditions that are difficult to be met in densely populated areas

Legend Pocket park, Group service units, Slow-traffic system, River system

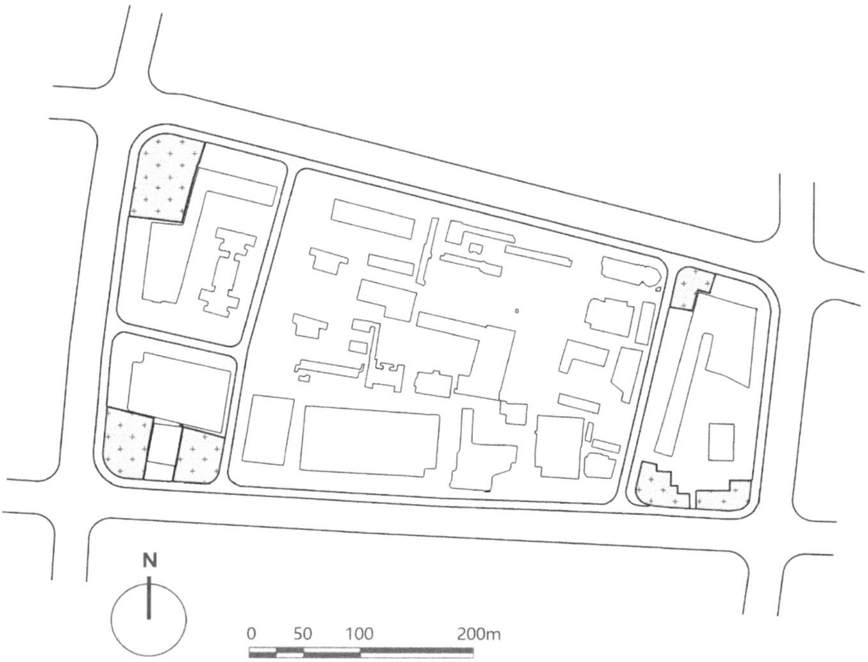

Fig. 4.21 Example of ring structure: Hongwu North Road Changjiang Road Southeast Block (Nanjing, China)

boundary areas (Fig. 4.21). In this structure, pocket parks can establish a continuous and enclosed service sequence, facilitating activities such as daily walking, running, cycling, dog walking, and so on. However, the main disadvantage of this structure is the relatively singular links between parks and the simplicity of their interaction patterns.

For the groups in radial structure, it is commonly found that some pocket parks are situated at the center of the service unit, while others are located on its periphery. The pocket parks locating in the unit center and periphery are linked through a radiative slow-traffic system (Fig. 4.22). This structure is usually observed in the neighborhoods with central open spaces. It can effectively concentrate the service capacity of a pocket park group into the internal area of the unit, thereby enhancing the park accessibility of the areas where the residents gather together. However, it is not easy to form a coherent and closed park service sequence in this structure. Moreover, the great discrepancies in service efficiency between central and peripheral pocket parks would easily lead to imbalances in service effectiveness of the whole group.

In linear structure, pocket parks are typically positioned at one or two edges of the service unit or distributed along an internal open space strip. This structure is commonly found in the neighborhoods with limited internal open spaces or with limited access (Fig. 4.25). Pocket parks in this structure can form a continuous

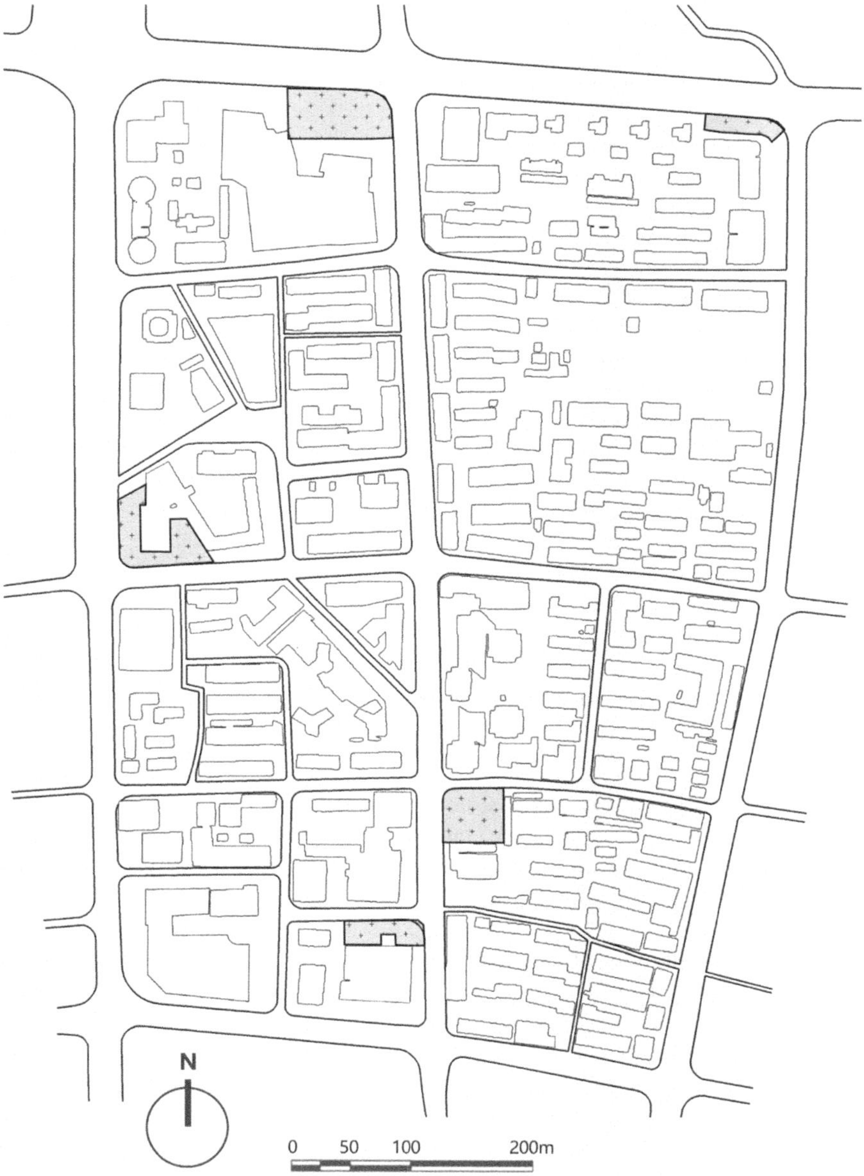

Fig. 4.22 Example of radial structure: Zhongshan Road Zhujiang Road Northeast Block (Nanjing, China)

and non-closed service sequence, reserving extension interfaces to connect with the green space system of the adjacent neighborhoods or blocks (Fig. 4.23). However, the parks in linear structure often deviate from the center of the service unit, resulting in significant discrepancies in park accessibility of the residents in different areas within the unit. Furthermore, similar to the groups in ring structure, the connection patterns between the pocket parks are also relatively limited.

For the groups of network structure, pocket parks are dispersedly distributed throughout the service unit and linked by a slow-traffic network (Fig. 4.24). This structure usually exists in open neighborhoods with ample internal open spaces and a well-developed slow-traffic system. Facilitated by the advanced linkage network, the pocket parks possess the capability of establishing diverse service sequences and multiple circulation systems, bringing high accessibility for all the residents within the unit. However, the formation and construction of the network structure usually necessitate more rigorous conditions, which is specially challenging in most densely populated areas.

Fig. 4.23 Example of linear structure: Taiping South Road Zhongshan East Road Southeast Block (Nanjing, China)

Fig. 4.24 Example of network structure: Taiping North Road Zhujiang Road Southeast Block (Nanjing, China)

In the planning process, as no spatial structure holds absolute superiority or inferiority, there is no "one-size-fit-all" structure for all planning scenarios. The selection and formation of spatial structure greatly hinges on the joint influences of the factors like land use structure, development intensity, urban fabric, and transportation system. The capability of establishing an appropriate spatial structure for the service units directly influences the service effectiveness of pocket park groups and their response degree to recreational demands, which will be extensively discussed in Chapter 5.

(2) Measurement

The measurement of the spatial structure of the group mainly focuses on quantifying its absolute and relative spatial characteristics. The absolute spatial characteristics primarily reflect the objective spatial attributes of the structure, including the group size, density, average distance between pocket parks, average number of park links, and so on [36, 37]. The relative spatial characteristics delineate the

Fig. 4.25 Application of spatial structure metrics

group morphology, the degree of agglomeration, and other characteristics, which are commonly measured by the metrics like the morphological compactness and the degree of distributional agglomeration [38]. According to the application in planning practice and research, an integrated application of these two types of indicators can unfold a full picture of the spatial relationship among the parks within the group (Table 4.6, Fig. 4.25).

4.4 Collaboration

4.4.1 Group Collaboration

In a system like the pocket park group, its element collaboration is seemed as a result of synergy. The concept of "synergy" originated in physics and German physicist Hermann Haken developed it into a theory known as "Synergistics" in the 1970s

Table 4.6 Metrics for spatial structure measurement

Item	Metrics	Formula	Description
The absolute spatial characteristics	Density	DenGr = k/s DenGr is the density of pocket parks, k is the number of pocket park in the group, s is the size of the service unit	The higher the value, the higher the density of pocket parks in the service unit
	Average distance between pocket park and its nearest pocket park	$\overline{D}_o = \sum_1^k D_i/k$ $\overline{D}_o$ is the average distance between pocket park and its nearest pocket park, k is the number of pocket park in the group, D_i is the path distance between pocket park i and its nearest pocket park	The lower the value, the greater the likelihood of interaction among the pocket parks within the group
	Average number of links	$AveNum = \sum_1^k n_i/k$ AveNum is the average number of links, k is the number of pocket parks in the group, n_i is the number of pocket parks in the groups that have a direct link to the pocket park i	The higher the value, the greater the complexity of the linkage network within the group
The relative spatial characteristics	Layout form compactness	$LFCR = \sqrt{\pi a}/p$ LFCR is the layout form compactness rate, a is the convex polygon area of the pocket park groups, ǁ is the perimeter of the convex polygon of groups	The higher the value, the more compact the spatial distribution of the pocket parks within the group
	Average nearest neighborhood index	$ANNR = \overline{D}_O/\overline{D}_E$ ANNR is the average nearest neighborhood index, $\overline{D}_O$ is the average distance of the closest neighboring pocket park of the park group, $\overline{D}_E$ is the average distance in condition that the pocket parks are randomly distributed	When the value is less than 1, the observed pattern is clustering, whereas a value more than 1 indicates a tendency towards a discrete pattern

[39]. Synergy refers to the coordination and collaboration of the internal subsystems or elements of the complex open system. In this process, the internal subsystems or elements interact with each other to create an organized spatial, temporal, or functional structure [39]. The term "Synergistic Effect" describes the overall or collective impact resulting from the interaction of numerous subsystems within an open system, often illustrated by the "1 + 1 > 2" phenomenon [40]. Analytical models grounded in synergy theory are employed to qualitatively describe and quantitatively analyze the self-organization states across diverse systems [41]. Synergistics elucidates the general principles and laws of the system self-organization, with applications spanning various disciplines of natural [42] and social science [43].

Urban green space system exhibits all the characteristics of open system, which can be an ideal object for the application of synergy theory. This application includes the examination of the interrelationships of the growth and spatial distribution between urban green spaces and residents [44], the overall response level to the various demands [45], and the co-development degree of the urban green spaces and other related elements [46, 47]. In contrast to previous studies focusing on the collaboration between urban green spaces and external service objects or environmental elements, our study on pocket park groups emphasizes the interactions among parks. The core objective is to clarify the interaction mechanism of the constituents (pocket park individuals) of the park group in the recreational service process, which can offer valuable insights to the planning and design practice of pocket parks.

Integrating the connotation of synergy with the service attributes of pocket parks, the collaboration of pocket park group can be characterized as the comprehensive effect generated by the mutual coordination. In this process, the collaboration relationship is primarily measured through the service structure indicators (Table 4.4), while the collaboration results are chiefly quantified by the service performance indicators (Table 4.7).

4.4.2 Group Performance

The concept of "Performance" was initially introduced from the field of management studies, reflecting people's achievements or results of a specific activity [48]. It was subsequently employed as an indicator and management strategy to enhance enterprise productivity. By the late twentieth century, "performance" started gaining traction in urban planning and was used to evaluate the running status of the specific facets of urban system, encompassing economic [49], environmental [50], and social [51] dimensions. Moreover, its application also extends to the evaluation on the service status of specific land uses and facilities, such as industrial zones [52], commercial zones [53], and transportation facilities [54]. The performance of urban green spaces falls into the latter application mode and is mainly used to measure and describe the service status of green spaces. Accordingly, the performance of a pocket park group reflects the collaboration result of the group members and represents their total and collective service status. To measure this, indicators can be classified into the categories of effectiveness and efficiency.

(1) Effectiveness

According to the core value and service characteristics of pocket parks, the effectiveness indicators should focus on the measurement of the visitor volume, shareability (the degree of sharing of different user groups), and inclusiveness (the degree of compatibility of different activities). Of these three aspects, the visitor volume of pocket park group can be quantified by the sum of all parks' visits, while the shareability and inclusiveness can be both measured using the diversity indicators, such as the Shannon–Wiener Index and Simpson Index. Diversity indicators were originally employed in ecology to assess biodiversity [55]. The Shannon–Wiener Index uses the uncertainty measurement principle from information theory to predict the variations

Table 4.7 Metrics for pocket parks groups' service performance measurement

Type	Item	Metrics	Formula	Description
Effectiveness	Visitor volume	The sum of the visits of all the parks in the group	$N = \sum_1^K N_I$ N is the sum of the park visits of all the parks, k is the number of pocket park in the group, N_i is the visits of park i	The higher the value, the higher the visitor volume of the pocket park group
	Shareability	Shannon–Wiener Index of user groups	$H_C = -\sum_1^j G_i \mathrm{Ln}(G_i)$ Hc is the Shannon–Wiener Index of user groups, j is the number of the user group types, Gi is the proportion of user group i to the total number of visitors	The higher the value, the higher the shareability of the park spaces to the user groups
		Simpson Index of user group	$\mathrm{Sc} = 1 - \sum_1^j G_i^2$ Sc is the Simpson Index of user groups, j is the number of user group types, Gi is the proportion of user group i to the total number of visitors	The higher the value, the higher the shareability of the park spaces to the user groups
	Inclusiveness	Shannon–Wiener Index of activity types	$\mathrm{Hm} = -\sum_1^m O_i \mathrm{Ln}(O_i)$Hm is the Shannon–Wiener Index of activity types, m is the number of activity types, O_i is the proportion of the number of visitors doing activity i to the total number of visitors	The higher the value, the higher inclusiveness of the park spaces to the activity types
		Simpson Index of activity types	$\mathrm{Sm} = 1 - \sum_1^m O_i^2$ Sm is the Simpson Index of activity types, m is the number of activity type, O_i is the proportion of the number of visitors doing activity i to the total number of visitors	The higher the value, the higher the inclusiveness of the park spaces to the activity types
Efficiency	Space use efficiency	Density	$Denvi = N/S$ *Denvi* is the density of visitors, *N* is the total number of visitors of the park group, *S* is the total area of the pocket parks of the group	The higher the value, the higher the space use efficiency
		Per capita duration of park visitors	$\mathrm{AveTi} = \sum_1^N T_I/N$ AveTi is the per capita duration of park visitors, N is the total number of visitors of the park group, Ti is the duration of visitor i	The higher the value, the higher the space use efficiency
	Responsiveness	Park visit ratio	$R = N/P$R is the park visit ratio of the unit's residents, N is the total number of visitors of the park group, P is the total population of the service unit	The higher the value, the higher the responsiveness

Note In practice, the Shannon-Weiner Index and Simpson's Index have similar indications for selection. Therefore, it is sufficient to select either one of them in use in the number of species that an individual may encounter within a specific community [56]. Higher diversity within a community corresponds to increased uncertainty. Meanwhile, the measurement principle of Simpson Index is to assess the likelihood that two randomly sampled individuals belong to different species [57]. Both indicators account for the number of types (such as user group types and activity types) and the evenness of individual distribution among them (user group evenness or activity evenness). Higher index values signify greater diversity of user groups or activity types, as well as increased shareability and inclusiveness of the pocket parks.

(2) Efficiency

Conventional indicators employed to assess the urban green space efficiency typically include density and per capita visit time of park visitors. However, the size difference between the small (Paley Park with 0.04 hectare) and large pocket park (1 hectare in China) can reach to more than 20 times, which probably lead to significant fluctuation of the visitor density in different parks, thus compromising their applicability and reliability. Additionally, measuring the per capita visit time is also a challenging task since the pocket park visitors are highly fluid and dynamic whose visit time is commonly short compared to large park visitors. Correspondingly, the park visit ratio and mean weekly visit frequency of residents of the service unit can be served as additional efficiency indicators, reflecting the responsiveness of park group to residents' demands. When comparing service performance across different pocket park groups, discrepancies between effectiveness and efficiency should prompt comprehensive evaluation rather than relying solely on the value of one type of indicators.

4.4.3 Collaborative Mode

The purpose of group collaboration is to enhance their overall recreation attractiveness and service quality, ultimately optimizing the group's overall performance. Taking the total visitor volume (the sum of park visits) as the main indicator, we carried out a field survey and statistical analysis on pocket park groups in Nanjing. Then, we summarized the collaborative modes of park groups, mainly the groups showing relatively high visitor volume, by examining their service structure over the entire period and subperiods in the survey (Table 4.8):

(1) General Collaboration

Coupling with the types of service structure, the collaboration of park groups during the entire period (7:00–21:00) can be classified into center-based and no-center-based mode. The center-based collaboration proves prominent service center(s) while these groups commonly show a vertical division structure (Fig. 4.26). Contrarily, the service performance of individual pocket parks in the no-center-based collaboration is relatively balanced, most of which employ the horizontal division structure in service (Fig. 4.27).

Table 4.8 Characteristics of collaborative mode

Time period	Collaborative mode	Characteristics
General collaboration	Center-based	• Clear service centers • Vertical division structure
	No-center-based	• Relatively balanced service performance of individual pocket park • Greater complementarity among parks • Horizontal division structure
Instant collaboration	Consistent center-based	• Same service centers in all subperiods • Stable service performance of individual parks across the entire period • Center-based collaboration mode in general
	Variable center-based	• Different service centers in various subperiods • Varied service performance of individual pocket parks in subperiods • Either center-based or no-center-based collaboration mode in general
	No-center-based	• No obvious service center at any subperiod • Stable service performance of individual pocket parks across the entire period • No-center-based collaboration in general

(2) Instant Collaboration

A further analysis was made on the park visits of the sample groups in the seven subperiods of our survey and reveals three instant collaborative modes, including consistent center-based, variable center-based, and no-center-based. The consistent center-based collaboration indicates that the group service is continuously dominated by the same service center(s) in all the subperiods, which commonly forms the general collaboration mode of center-based (Fig. 4.28). The variable center-based collaboration denotes that the group service is dominated by different park centers in various subperiods, typically leads to either a center-based or no-center-based collaboration in general (Fig. 4.29). The no-center-based collaborative mode suggests that the group has no obvious service center park at any subperiod, usually leading to a no-center-based collaboration in the entire period.

4.4.4 Influences of Group Performance

Performance is a direct indication of group collaboration and can also be used as a tool for evaluating the collaborative degree and mode.

(1) Influences of Visitor Volume

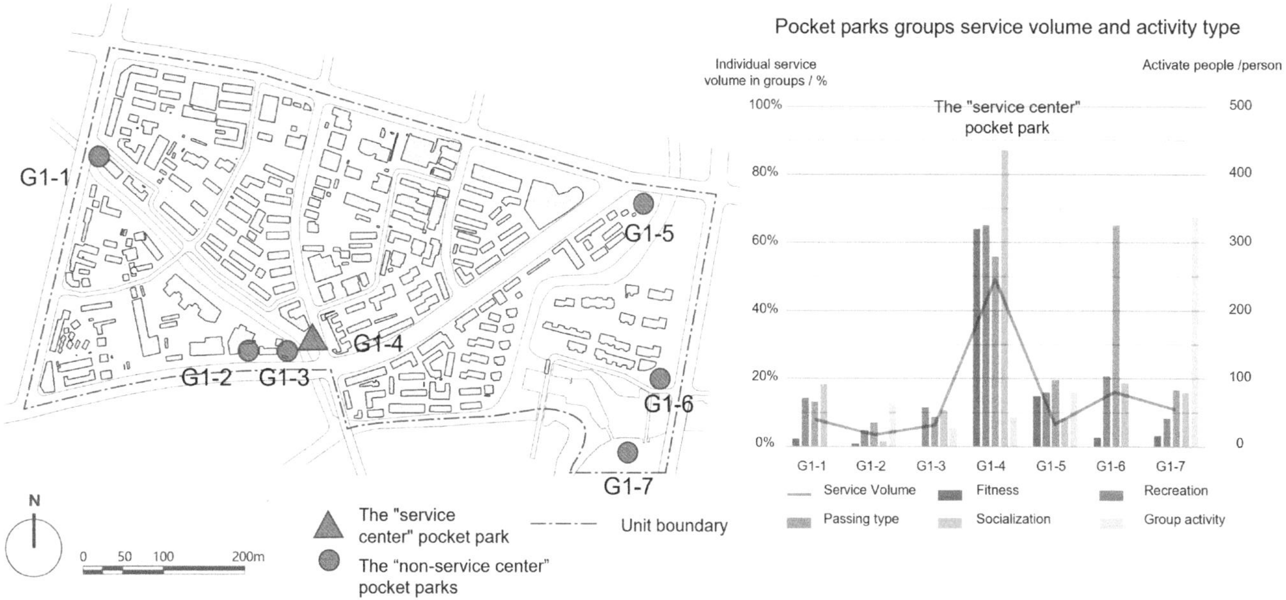

Fig. 4.26 Example of center-based general collaboration: Baixia Road Taiping South Road Southeast Block (Nanjing, China)

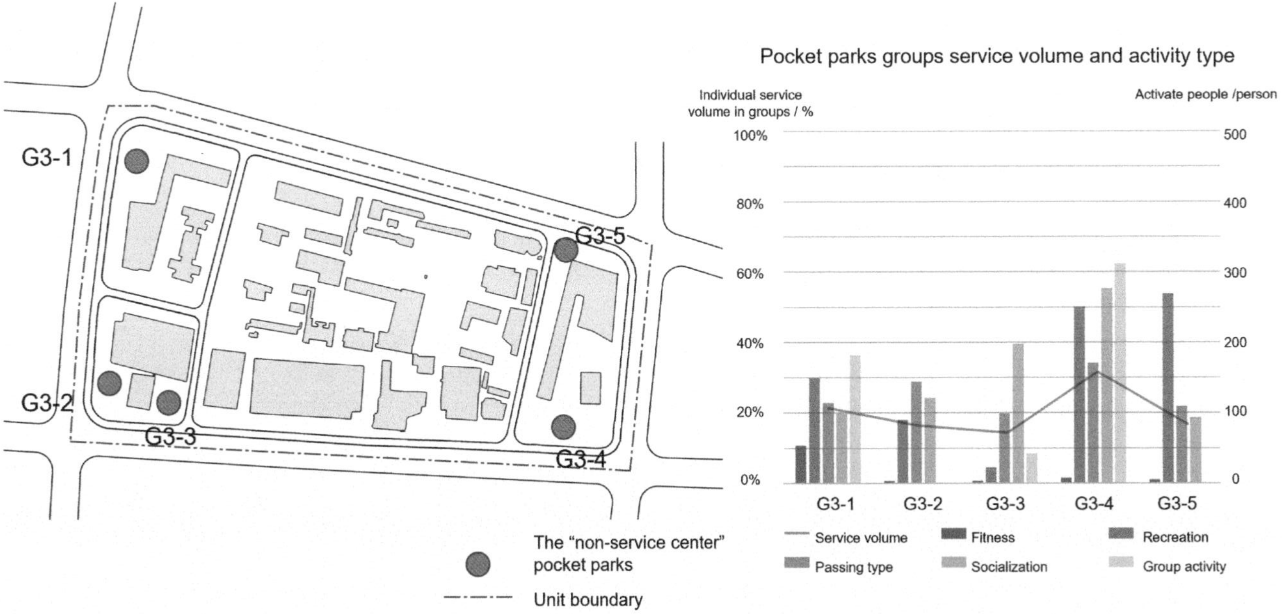

Fig. 4.27 Example of no-center-based general collaboration: Changjiang Road Hongwu North Road Southeast Block (Nanjing, China)

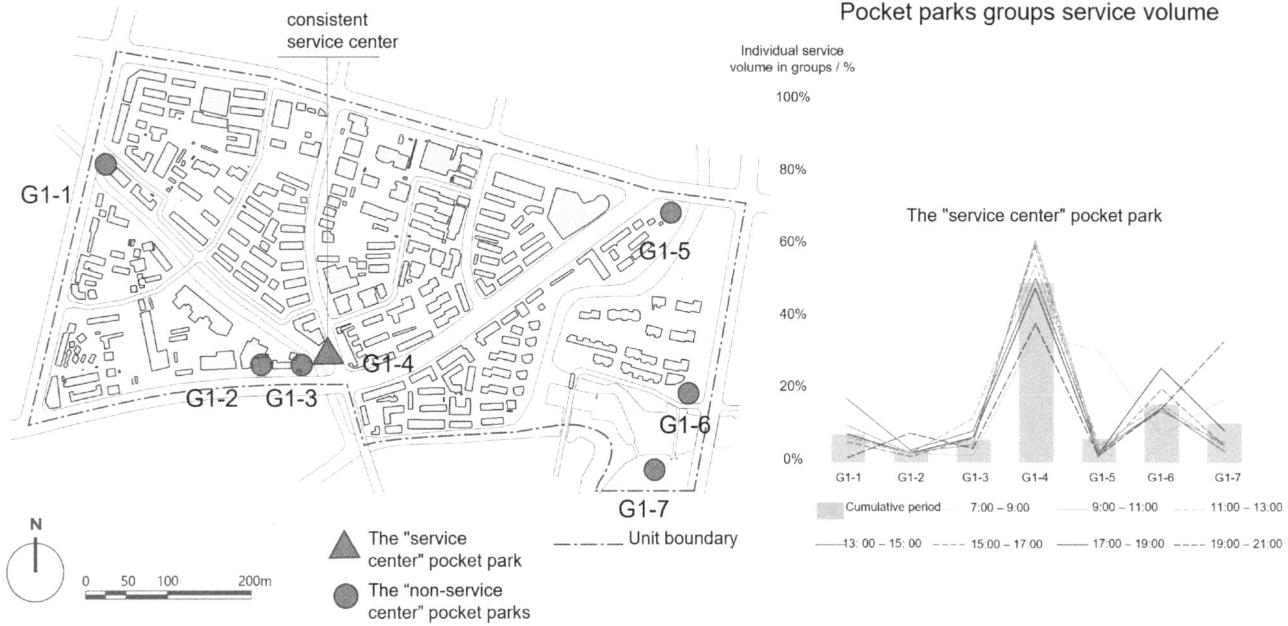

Fig. 4.28 Example of consistent center-based instant collaboration: South Taiping Road Baixia Road Southeast Block (Nanjing, China)

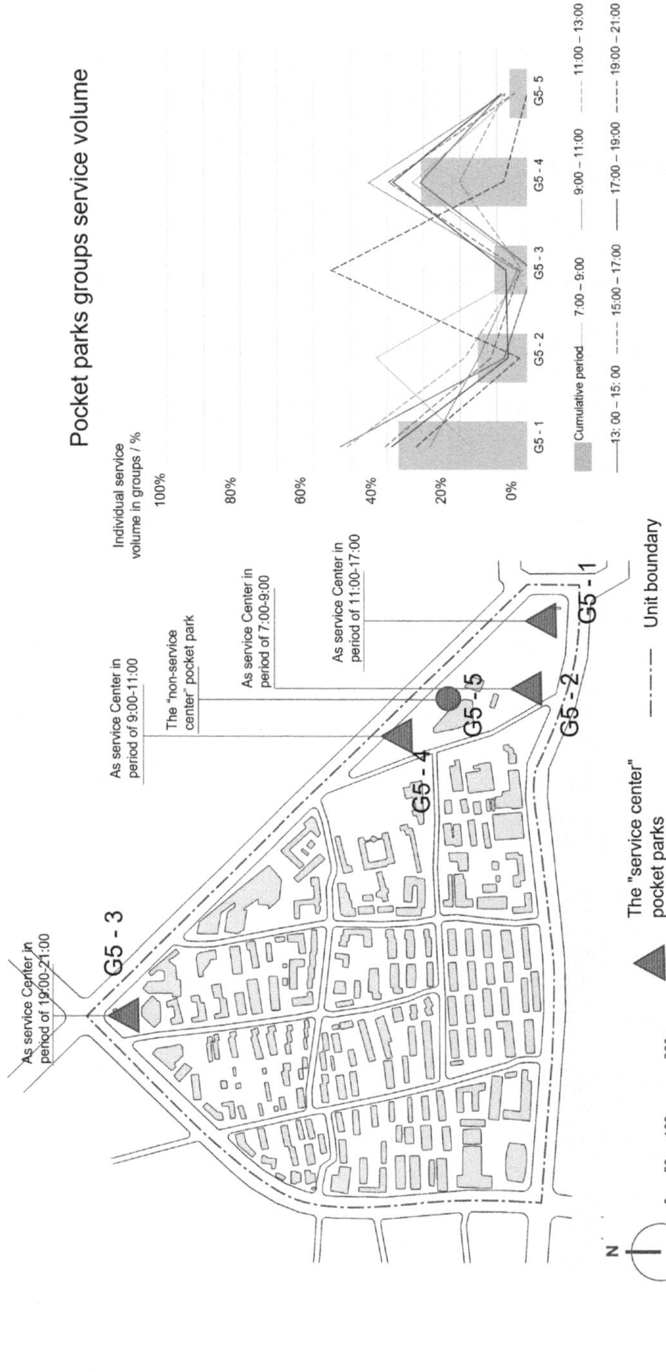

Fig. 4.29 Example of variable center-based instant collaboration: Yunnan Road West Beijing Road Northeast Block (Nanjing, China)

The visitor volume is proved to be tightly associated with the size and proximity of potential users [58]. For instance, pocket parks in the service unit with larger population sizes and higher proximity commonly show increased park visits. Secondly, parks' service capacity (e.g., total area, activity site area, etc.) also significantly affects park visits [59]. Particularly in the areas where the green space resource is scarce, the insufficient capacity of park group directly constrains total visits. Additionally, recreational facilities surrounding the pocket parks can enhance service quality and attractiveness, thereby boosting visit frequency [60].

Moreover, group characteristics, such as service and spatial structure, influence park visits [61]. Our study in Nanjing revealed that pocket park groups with diverse and flexible service division mode can better cater to the demands of various users, consequently increasing the park visits [62]. Likewise, more compact spatial structure and closed interconnection of parks facilitate the fostering of the synergetic effect within the groups, resulting in enhanced parks' mutual complementarity and supportiveness.

(2) Influences of Service Diversity

Our study in Nanjing also revealed that the diversity of user groups and their activities showed a high correlation, for which both of them can be classified as service diversity indicators (Fig. 4.30). The service diversity level directly corresponds to the composition of resident communities within the service unit. Greater diversity of the potential users raises higher demand for various services of pocket parks, increasing the likelihood of varied behaviors and activities.

Furthermore, the service division mode and spatial structure of pocket park group affect the inclusiveness of different user groups and activities. For example, presetting different dominant functions to the pocket parks within a group is conducive to establish a horizontal service division mode that can effectively accommodate the use of diverse groups. Similarly, a refined analysis of user groups' demands and activity characteristics can facilitate the establishment of a composite service division mode to offer staggered and flexible services. In addition, expanding the area and proportion of activity sites can help to maintain the capacity and usage flexibility of pocket parks in condition that the park size is severely limited.

(3) Influences of Demand Responsiveness

The degree of demand responsiveness serves as a comprehensive indicator encompassing the measurements of the conditions of both demand-side and supply-side. The fit degree between supply and demand conditions determines the extent of demand responsiveness [63]. This supply–demand fit entails consideration of both quantitative and qualitative aspects. Quantitatively, it focuses on aligning the size of pocket parks with the size of target demand to achieve the size fit. Qualitatively, it encompasses two dimensions: (1) pocket parks are supposed to provide the service fitting to the expected quality of demand side (such as accessibility, service type, vegetation, space quality, etc.); (2) the division and service mode of pocket park group should couple with the composition and demand structure of the user groups, thereby precisely providing whatever is requested.

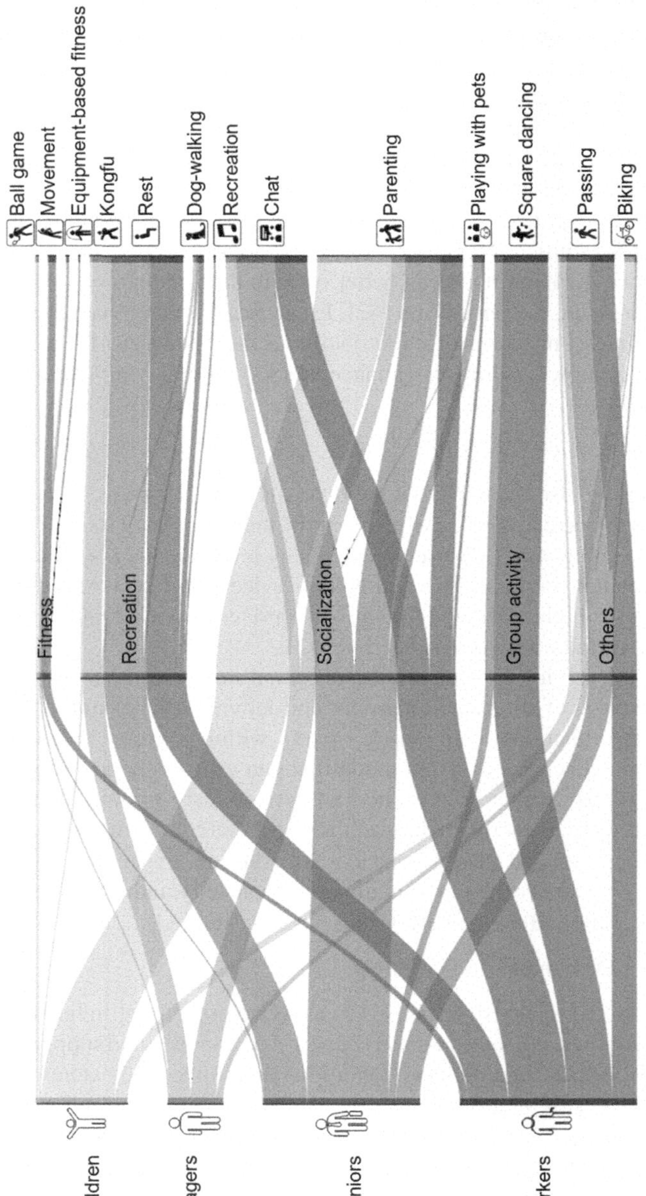

Fig. 4.30 Activity type composition of different social groups in a pocket park

References

1. Liu H, Remme RP, Hamel P et al (2020) Supply and demand assessment of urban recreation service and its implication for greenspace planning—a case study on Guangzhou. Landsc Urban Plan 203:103898. https://doi.org/10.1016/j.landurbplan.2020.103898
2. Gibson H, Canfield J (2016) Pocket parks as community building blocks: a focus on Stapleton, Co. Community Dev 47(5):732–745. https://doi.org/10.1080/15575330.2016.1220965
3. Zhou C, An Y, Zhao J et al (2022) How do mini-parks serve in groups? A visit analysis of mini-park groups in the neighbourhoods of Nanjing. Cities. Advance online publication. https://doi.org/10.1016/j.cities.2022.103804
4. Rosso F, Cappa F, Spitzmiller R et al (2021) Pocket parks towards more sustainable cities. Architectural, environmental, managerial and legal considerations towards an integrated framework: a case study in the Mediterranean region. Environmental Challenges. Advance online publication. https://doi.org/10.1016/j.envc.2021.100402
5. Ministry of Housing and Urban-Rural Development of the People's Republic of China (2018) Standard for urban residential area planning and design: GB50180-2018. China Architecture & Building Press, Beijing
6. Shanghai Urban Planning and Land Resources Administration Bureau. (2016). Shanghai planning guidance of 15-minute community-life circle
7. Mohai P, Pellow D, Roberts JT (2009) Environmental justice. In: Annual review of environment and resources (Vol 34, pp 405–430). Annual Reviews. https://doi.org/10.1146/annurev-environ-082508-094348
8. Dai D (2011) Racial/ethnic and socioeconomic disparities in urban green space accessibility: where to intervene? Landsc Urban Plan 102(4):234–244. https://doi.org/10.1016/j.landurbplan.2011.05.002
9. Boone CG, Buckley GL, Grove JM et al (2009) Parks and people: an environmental justice inquiry in Baltimore, Maryland. Ann Assoc Am Geogr 99(4):767–787. https://doi.org/10.1080/00045600903102949
10. Wüstemann H, Kalisch D, Kolbe J (2017) Access to urban green space and environmental inequalities in Germany. Landsc Urban Plan 164:124–131. https://doi.org/10.1016/j.landurbplan.2017.04.002
11. Kabisch N, Haase D (2014) Green justice or just green? Provision of urban green spaces in Berlin, Germany. Landsc Urban Plan 122:129–139. https://doi.org/10.1016/j.landurbplan.2013.11.016
12. Richards DR, Passy P, Oh RRY (2017) Impacts of population density and wealth on the quantity and structure of urban green space in tropical Southeast Asia. Landsc Urban Plan 157:553–560. https://doi.org/10.1016/j.landurbplan.2016.09.005
13. Tan PY, Samsudin R (2017) Effects of spatial scale on assessment of spatial equity of urban park provision. Landsc Urban Plan 158:139–154. https://doi.org/10.1016/j.landurbplan.2016.11.001
14. Denhardt JV (2007) The new public service: serving, not steering (Expanded edition). M.E. Sharpe, Armonk, NY. https://search.library.wisc.edu/catalog/9910032518202121
15. Levy R (2010) New public management: End of an era? Public Policy and Administration 25(2):234–240. https://doi.org/10.1177/0952076709357152
16. Zhou C (2020) Research on the evolution of the connotation of "equity" and its measurement standards in park land planning. Chinese Landsc Archit 36(300):52–56. https://doi.org/10.19775/j.cla.2020.12.0052
17. Zhou C (2014) The method for urban green space system planning formulation, 1st edn. Southeast University Press, Nanjing
18. Greater London Authority, City Hall, The Queen's Walk, et al (2008) The London Plan-The Spatial Development Strategy for Greater London. https://www.london.gov.uk/thelondonplan
19. The House of Representatives, Japan. (1956). Urban park act. https://elaws.e-gov.go.jp/document?lawid=331AC0000000079

20. Ministry of Housing and Urban-Rural Development of the People's Republic of China (2019) Standard for planning of urban green space: GB/T51346–2019. China Architecture & Building Press
21. Whyte WH, Jr (1980) The social life of small urban spaces. Conservation Foundation, Washington, DC. https://search.library.wisc.edu/catalog/999547142202121
22. Zion R (1969) Parks where the people are—the small midtown park. New York University Press, New York
23. Gottmann J (1957) Megalopolis or the urbanization of the northeastern seaboard. Econ Geogr 33:189–200
24. Lambooy JG (1998) Polynucleation and economic development: the Randstad. Eur Plan Stud 6(4):457–466. https://doi.org/10.1080/09654319808720474
25. Bailey N, Turok I (2001) Central Scotland as a polycentric urban region: useful planning concept or chimera? Urban Studies 38(4):697–715. https://doi.org/10.1080/00420980120035295
26. Halbert L (2008) Examining the mega-city-region hypothesis: evidence from the Paris city-region/Bassin parisien. Reg Stud 42(8):1147–1160. https://doi.org/10.1080/003434007018 61328
27. Hoffer CR (1931) Understanding the community. Am J Sociol 36(4):616–624. https://doi.org/ 10.1086/215480
28. Grant J (2005) Planning the good community new urbanism in theory and practice, 1st edn. Routledge, London
29. Shanghai Municipal Bureau of Planning and Natural Resources (2016) Shanghai's 15-minute community life circle planning guidelines
30. Johnson DL (2002) Origin of the neighbourhood unit. Plan Perspect 17(3):227–245. https:// doi.org/10.1080/02665430210129306
31. Perry CA (2011) The neighborhood unit. The city reader. https://api.semanticscholar.org/Cor pusID:169545535
32. Xie C, Wang J, Haase D et al (2023) Measuring spatio-temporal heterogeneity and interior characteristics of green spaces in urban neighborhoods: a new approach using gray level co-occurrence matrix. Sci Total Environ 855:158608. https://doi.org/10.1016/j.scitotenv.2022. 158608
33. Song Y, Merlin L, Rodriguez D (2013) Comparing measures of urban land use mix. Comput Environ Urban Syst 42:1–13. https://doi.org/10.1016/j.compenvurbsys.2013.08.001
34. Rhoades SA (1993) The Herfindahl-Hirschman index. Federal Reserve Bulletin, 188–189
35. Turner MG, Gardner RH (2015) Landscape ecology in theory and practice: pattern and process. Springer, New York, NY. https://doi.org/10.1007/978-1-4939-2794-4. https://doi.org/10.1007/ 978-1-4939-2794-4
36. Wu Y, Patuano A, Mashhoodi B et al (2025) How small green spaces cool urban neighbourhoods: optimising distribution, size and shape. Landsc Urban Plan 253:105224. https://doi.org/ 10.1016/j.landurbplan.2024.105224
37. Park J, Kim JH, Lee DK et al (2017) The influence of small green space type and structure at the street level on urban heat island mitigation. Urban Forestry & Urban Greening 21:203–212. https://doi.org/10.1016/j.ufug.2016.12.005
38. Chen J, Kinoshita T, Li H et al (2024) Which green is more equitable? a study of urban green space equity based on morphological spatial patterns. Urban Forestry & Urban Greening 91:128178. https://doi.org/10.1016/j.ufug.2023.128178
39. Haken H (2004) Synergetics: Introduction and advanced topics (Vol 1, 1st ed) Springer Berlin, Heidelberg. https://doi.org/10.1007/978-3-662-10184-1
40. Geary N (2013) Understanding synergy. Am J Physiol Endocrin Metabol 304(3):E237–E253. https://doi.org/10.1152/ajpendo.00308.2012
41. Kröger B (2015) Theories of self-organization: the role of synergetics. In: Kröger B (ed) Hermann Haken: from the laser to synergetics—a scientific biography of the early years. Springer International Publishing, Cham, pp 211–227. https://doi.org/10.1007/978-3-319-116 89-1_9

42. Corning PA (2014) Systems theory and the role of synergy in the evolution of living systems. Syst Res Behav Sci 31(2):181–196. https://doi.org/10.1002/sres.2191
43. George AS, Scott K, Mehra V et al (2016) Synergies, strengths and challenges: findings on community capability from a systematic health systems research literature review. BMC Health Serv Res 16(7):623. https://doi.org/10.1186/s12913-016-1860-1
44. Kienast F, Degenhardt B, Weilenmann B et al (2012) GIS-assisted mapping of landscape suitability for nearby recreation. Landsc Urban Plan 105(4):385–399. https://doi.org/10.1016/j.landurbplan.2012.01.015
45. Meerow S, Newell JP (2017) Spatial planning for multifunctional green infrastructure: growing resilience in Detroit. Landsc Urban Plan 159:62–75
46. Liu B, He W, Liu S (2012) Study of the evaluation and planning of urban green space based on the coupling theory of green space and city space. Chinese Landscape Architecture 28(05):42–46
47. Liu S, Liu B (2010) Coupling analysis between urban green space and urban development: a case study of Wuxi. Chinese Landscape Architecture 26(3):14–18
48. Stredwick J (2013) An introduction to human resource management, 3rd edn. Routledge, London
49. Succurro M (2020) Economic performance. In: Marciano A, Ramello GB (eds) Encyclopedia of law and economics. Springer, New York, NY, pp 1–4. https://doi.org/10.1007/978-1-4614-7883-6_208-2
50. Song ML, Fisher R, Wang JL et al (2018) Environmental performance evaluation with big data: theories and methods. Ann Oper Res 270(1):459–472. https://doi.org/10.1007/s10479-016-2158-8
51. Wood DJ (2010) Measuring corporate social performance: a review. Int J Manag Rev 12(1):50–84. https://doi.org/10.1111/j.1468-2370.2009.00274.x
52. Xie H, Wang W, Yang Z et al (2016) Measuring the sustainable performance of industrial land utilization in major industrial zones of China. Technol Forecast Soc Chang 112:207–219. https://doi.org/10.1016/j.techfore.2016.06.016
53. Singh K, Hachem-Vermette C (2019) Impact of commercial land mixture on energy and environmental performance of mixed use neighbourhoods. Build Environ 154:182–199. https://doi.org/10.1016/j.buildenv.2019.03.016
54. Raoniar R, Rao M, Senathipathi V (2015) Public transport performance evaluation techniques—a review, 3:1–8
55. Roberts FS (2019) Measurement of biodiversity: richness and evenness. In: Kaper HG, Roberts FS (eds), Mathematics of planet earth: protecting our planet, learning from the past, safeguarding for the future. Springer International Publishing, Cham, pp 203–224. https://doi.org/10.1007/978-3-030-22044-0_8
56. Strong WL (2016) Biased richness and evenness relationships within Shannon-Wiener index values. Ecol Ind 67:703–713. https://doi.org/10.1016/j.ecolind.2016.03.043
57. Feranchuk S, Belkova N, Potapova U et al (2018) Evaluating the use of diversity indices to distinguish between microbial communities with different traits. Res Microbiol 169(4):254–261. https://doi.org/10.1016/j.resmic.2018.03.004
58. Donahue ML, Keeler BL, Wood SA et al (2018) Using social media to understand drivers of urban park visitation in the Twin Cities, MN. Landsc Urban Plan 175:1–10. https://doi.org/10.1016/j.landurbplan.2018.02.006
59. Geng H, Zhang Y, Chi J, et al (2024) What affect the satisfaction, preferences, and visitation of pocket parks? Evidence from Shanghai. J Outdoor Recreat Tour. Advance online publication. https://doi.org/10.1016/j.jort.2024.100764
60. Huai S, Liu S, Zheng T et al (2023) Are social media data and survey data consistent in measuring park visitation, park satisfaction, and their influencing factors? a case study in Shanghai. Urban Forest Urban Green 81:127869. https://doi.org/10.1016/j.ufug.2023.127869
61. Qiu F, Hu X (2010) Modeling group structures in pedestrian crowd simulation. Simul Model Pract Theory 18(2):190–205. https://doi.org/10.1016/j.simpat.2009.10.005

62. Zhou C, Xie M, Zhao J, et al (2022) What affects the use flexibility of pocket parks? Evidence from Nanjing, China. Land. Advance online publication. https://doi.org/10.3390/land11091419
63. Xian L, Yin H, Fang J (2023) Evaluating the supply-demand relationship for urban green parks in Beijing from an ecosystem service flow perspective. Urban Forest Urban Green 85:127974. https://doi.org/10.1016/j.ufug.2023.127974

Chapter 5
Site Selection and Service Provision

The main tasks of green space planning are tightly associated with the issues and challenges that need to be handled in cities' specific context, which are dynamic in different spatiotemporal conditions of urban development [1]. Unlike the holistic construction of the green space system during fast urban expansion, the development of pocket parks is to continuously provide new green spaces to improve the existing service, thereby creating incremental social and ecological values for the city [2]. As the core task of planning, pocket parks' site selection and service provision greatly determine the effectiveness of pocket park development, which includes the responsiveness to people's demands [3], allocational equity of public resources [4], suitability to the existing environments, and group collaborative efficiency [5].

Since this task predominantly occurs within the built environment where the main framework of green space system and other infrastructure system have already been built up, it is essential to carefully examine this task in city's existing conditions. By this way, the installation of pocket parks can sufficiently leverage the present service offered by the existing facilities and infrastructures, thereby compensating for pocket parks' service deficiency and saving the construction cost [6]. Furthermore, numerous essential information supporting the decision-making of pocket parks' site selection and service provision also lies in the built environment, such as the supply–demand status of recreational resources [7], land acquisition opportunities [8], and construction and implementation suitability[8, 9], which further raises high requirements on the breadth, precision, and granularity of the measurement of the existing condition. To facilitate this examination, we discussed the measurement criteria, framework, and methods in this chapter.

C. Zhou, *Principles and Methods of Pocket Park Planning and Design*, Urban Sustainability, https://doi.org/10.1007/978-981-96-5801-5_5

5.1 Criteria of Effectiveness

5.1.1 Responsiveness

As practicality is one of the core values of pocket parks, pocket parks' site selection and service provision should prioritize the responsiveness to residents' recreational demands. Meanwhile, high responsiveness to demands is also essential for maximizing the service performance of both individual pocket park and park group. To achieve this goal, it is necessary to couple the pocket park distribution with residents' demand distribution from three aspects: location, quantity, and service structure.

First, coupling the location of parks and demands involves the work of the identification of demands' spatial distribution, which offers a basis for the proximate installation of pocket parks. Second, quantifying the intensity of residents' demands provides the primary basis for deriving the proper quantitative attributes (number, size, and capacity) of pocket parks. Third, in order to accurately provide the park service, clarifying the structure of the potential users and the composition of the demands of different user groups is essential, which can be used as the basis for service division, spatial arrangement, and facility conturation of pocket park groups.

5.1.2 Equity

According to the viewpoint of J. Rawls in *A Theory of Justice*, the essence of the equity issue in planning is to deal with the interest distribution of different groups in resource allocation [10]. The "environmental justice" is more specific and advocates that all social groups should have equal rights to enjoy public resources, including urban green spaces [11]. Therefore, if there is a problem of unequal distribution of public resources among various social groups, the redistribution of resources in planning should be prioritized to take care of the disadvantaged groups with less resource occupation. This approach is commonly called social relief, aiming to enhance the social equity and environmental justice of urban development [12, 13]. Significantly, one of the original goals of pocket park installation is to improve the severe imbalanced condition in green space resource allocation among different groups, especially in high-density cities. Following this logic, green space resources should be allocated preferentially to the places where the disadvantaged groups are located.

At present, equity measurement of green space planning is aggregated by four criteria, including the measurements of quantity, accessibility, group allocation, and demand responsiveness. The quantitative equity is mainly measured by the occupation indicators like parkland per capita, park number per capita, etc. [14]. Accessible equity is primarily measured through accessibility indicators, such as the number and area of parks that residents can reach within a 5-min walking distance and the rate of population covered by the park service radius [15, 16]. Group allocational equity is mainly measured by social equity indicators, such as the Gini coefficient, the Lorenz curve, etc. [17, 18]. Demand-responsive equity is primarily measured through the citizen satisfaction, which includes the responsiveness to the demands of each specific group on park quantity, accessibility, service facility, etc. [19, 20]. Correspondingly, the new installation of pocket parks should contribute to the enhancement of the aforementioned criteria.

5.1.3 Suitability

The suitability of the development environment and site is a critical factor in pocket parks' site selection and service provision, as it plays a key role in the cost control and parks' service performance. The suitability of the development environment primarily refers to the degree of support from the surrounding conditions to the park's service [8]. The determinants of the support degree include positive factors (such as recreation-related facilities, slow-traffic facilities, etc.) [3, 21]and negative factors (such as hazardous facilities, competing public parks, traffic barriers, etc.) [7, 22]. The more positive factors and fewer negative factors around the site, the more suitable it is to support the service of the pocket parks. Site suitability primarily refers to the appropriateness of the site conditions for installing pocket parks. The factors include the land acquisition opportunity [8], construction conditions (such as area, shape, slope, and surface attachments) [21], and so on. Typically, land with lower economic and social costs in acquisition and favorable construction conditions is more suitable for pocket park installation.

5.1.4 Collaboration

Existing studies showed that the proximately-located green spaces may produce both collaborative effects that mutually complement their service and competitive effects that mutually inhibit their service [5]. Therefore, maximizing the collaborative effects among green spaces is crucial for guaranteeing their service performance when grouping the pocket parks within a service unit. To achieve this goal, it involves meticulous guidance and control over the site selection and service definition of each pocket park within a group.

The collaboration of pocket parks within a group refers to the service complementarity and spatial correlation [23]. Service complementarity means that each pocket park within a group offers distinct services through space arrangement and facility configuration that cater to the unique demands of surrounding residents. This can provide residents with diverse choices, enhancing their recreational experiences and mitigating the homogeneous service provision and competition among parks. Spatial correlation indicates that pocket parks within a group are spatially interconnected by footpaths, bicycle lanes, comprehensive greenways, etc. and maintain a proper mutual distance that can guarantee the effective complementary support in service provision.

5.2 Technical Framework

5.2.1 Guidelines

In the early development of modern cities, green space planning always lagged behind the planning of residential lands, industrial lands, and other land uses, since green spaces cannot directly produce profit and contribute to city's economic growth, for which their development lands should be guaranteed by government intervention [24]. Accordingly, early urban green space planning usually adopted the planning mode dominated by the supply side which focused on the baseline guarantee of land use quota [25]. It means that the government mandates the threshold quota for green space development to guarantee the equivalent development opportunity of green spaces as the other land use categories.

In the subsequent urban development and renewal process, planning practices began to prioritize and proceed the concepts of people-oriented ideology[26], environmental justice [27], intensive land use [28], and so on, by which demand-side factors (e.g., accessibility, equity, etc.) gradually played an increasingly critical role in the allocation of public resources and became the key factors for decision-making in the planning of urban green spaces. Nowadays, with the completion or slowdown of the holistic urban renewal, cities are entering into the development phase focusing on micro-renewal and fine governance. The theory of "New Public Service", which is developed based on the democratic citizenship theory and community and civil society models, has become the principal guidance for urban public management [29]. Fulfilling the diverse demands and preferences of various resident groups has become an essential goal of planning [30]. To realize these, the approaches like city repair [31], city micro-surgery [32], and micro-reconstruction [33] can be increasingly observed, which also brings important opportunities for promoting the upgrading of the planning methods and technologies of pocket parks.

Throughout the development of green space planning, the dominant ideology has evolved from the early supply—oriented mode focusing on baseline guarantee of lands into the demand-oriented mode focusing on supply–demand fit. This shift has led to a continuous refinement of spatial and social granularity of the planning objects. The refinement of spatial granularity is evidenced by the shift of the green space planning unit, which converts from the previous large unit of the entire city or district into the present small unit of community or block. Similarly, the main objects of the planning also convert from the large-sized green spaces (like comprehensive parks and community parks) into the micro green spaces represented by pocket parks. Meanwhile, the refinement of social granularity is reflected by the subdivision of social attributes (such as age, gender, ethnicity, beliefs, income, etc.) of parks' target service groups, through which the demand to be fulfilled is being continuously diversified and specified. All these conversions jointly promote the transformation of the planning approaches of urban green spaces, from system construction mode into system optimization mode.

5.2.2 Determinants

Both the factors of supply-side and demand-side influence the balance process and results of pocket parks' site selection and service provision. Therefore, in order to achieve the goal of a seamless fit between supply and demand, it is essential to comprehensively measure and compare the factors of both sides that influencing the effectiveness of the site selection and service provision. Subsequently, the area with a high fit degree is preferred as the alternative site for pocket park installations.

The demand-side factors primarily refer to the primary service objects of pocket parks (i.e. residents) [22, 34]. Meanwhile, the supply-side factors primarily denote the conditions of the city or a specific area in supporting the pocket park development, which can be divided into the conditions of land use and environment [35, 36]. Land use conditions refer to the convenience in land acquisition for pocket park development, which is commonly determined by the current level of urban development and the maturity of the relevant policies [37]. For instance, in rapid urban expansion, green space development primarily relies on incremental development lands. Conversely, once urban expansion halts and enters into the phase of renewal, the opportunities for green space development should be exploited from the existing land resources. The land acquisition of pocket parks belongs to the latter way, which needs to search for development lands in the built environment, such as the existing leftover lots, vacant lots, and so on. Environmental conditions include the positive and negative impacts of the surrounding factors (such as surrounding land use, recreation-related facilities, and transportation network) on the construction, service, management, and maintenance of pocket parks [8, 36, 38].

5.2.3 Framework

Of the aforementioned factors, some can be measured by quantitative methods, such as the intensity of service demand, the suitability of the existing environment, etc. [22, 36]. However, there are still some factors that need to be analyzed through qualitative methods or integrally analyzed by both quantitative and qualitative methods. For example, the factors like policy, economy, and society, that constrain the development of pocket parks, are difficult to be measured by merely quantitative methods [39].

Correspondingly, a two-step strategy is recommended to achieve a relatively precise comparison and coupling analysis of the factors of supply and demand sides in the process of fixing the site and service of the pocket parks, which focuses on a sequential analysis of the quantifiable and non-quantifiable factors. First, it is proposed to quantitatively compare the factors of demand distribution and intensity on the demand side with the quantifiable factors of the environment conditions on the supply side. The sites with high fit degree of supply–demand conditions are extracted as the priority sites for pocket park installation. Second, qualitative analyses were made in conjunction with policy, economic, and social factors to ultimately determine the final sites and services of the parks (Fig. 5.1).

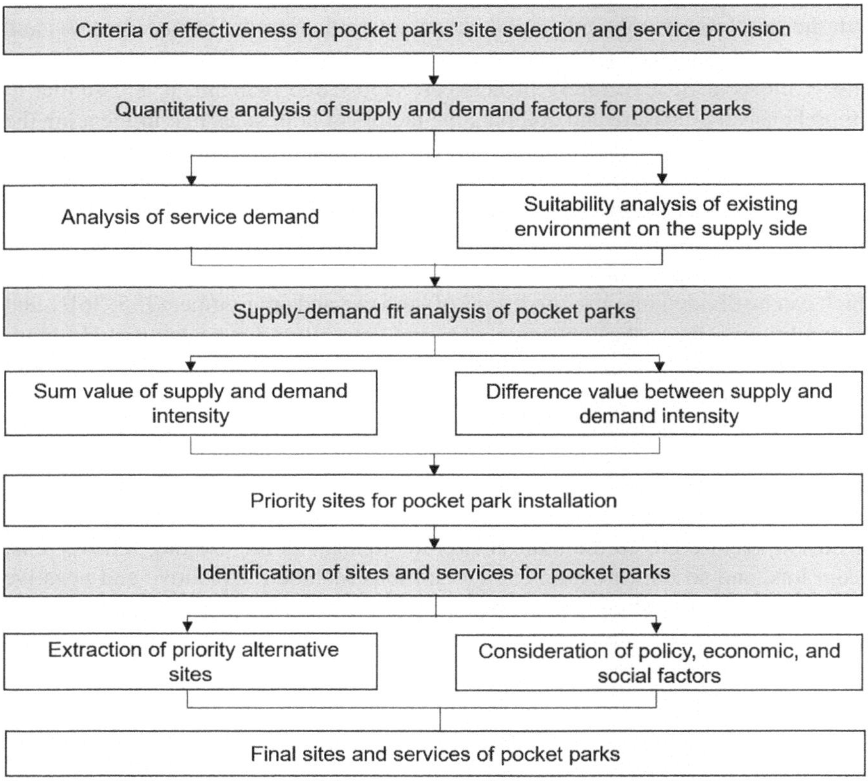

Fig. 5.1 Framework of pocket parks' site selection and service provision

5.3 Supply–Demand Fit

5.3.1 Factors of Demand-Side

As pocket parks' service primarily respond to daily recreational demands of the surrounding residents, the analysis of demand-side should highlight the distribution, density, and composition of residents. Since the target sites for pocket park installation concentrate in built-up areas where large-scale construction or renewal are basically completed, the population of the planning area seldom greatly fluctuates in a short term. Under this condition, the classification and analysis of residents' recreational demands in planning can be made relying on the population data of the existing conditions. Compared to the traditional planning that analyzes the population in a brief or coarse mode, the analysis for pocket park planning necessitates a refinement of the spatial and social granularity of the demand-side factors in order to precisely respond to the resident's demand.

(1) Fine-grained spatial units

Traditional urban green space planning typically makes the planning arrangement and indicator control focusing on the units of the entire city or its sub-districts. In contrast, the planning of pocket parks set the service unit of a pocket park group as the basic unit in planning control and regulation, raising higher requirements in precision due to the refinement of the spatial granularity of planning units. As mentioned in Chapter 4, the scope of the service units of pocket park groups, which are equal to the planning units, are commonly defined by two factors in space: community boundaries and trunk roads. By either of the two ways, the defined size of each unit is equivalent to 5–10 min life-circle neighborhood.

In high-density areas, the scope of the planning units defined by the two ways is quite similar, as most units' boundaries generated by these two ways coincide. Therefore, in practice, we can use one method to firstly delineate the scope of the units and use the other one to correct, optimize, and test the rationality of the preliminarily-delineated boundaries.

(2) Fine-grained social groups

In order to refine the social granularity, it requires a subdivision of the resident groups of the planning units according to their social attributes. Existing studies showed that the disparities of the social attributes, such as age, gender, income, profession, religion, ethnicity, etc., may affect the group's demand intensity and their usage pattern of green spaces [40]. Correspondingly, the refinement of the social granularity aims to clarify the intensity and types of the recreational demands of each group.

In particular, understanding the disparities of the demand intensity among different groups is crucial for determining the optimal quantity and locations for pocket park installation. For instance, using sampling survey to ascertain weekly expectation time of outdoor recreation activities across different age groups enables calculating the relative level of the recreational demand intensity of different groups [39]. Integrating the relative level of groups' recreational demand intensity with the demographic data of group composition can derive the total level of the recreational demand intensity of each community or other forms of planning unit [39].

The demand types of different social groups are commonly figured out through the scenario analysis or sampling survey. Observation can be used to analyze the utilization status of the built green spaces, revealing the demand and behavior disparities influenced by the identifiable attributes such as gender, race, and age. For instance, Danford [41] used an on-site observation method to analyze the usage patterns of people with different genders, skin colors, and ages in the pocket parks of Boston in pocket parks and confirmed the disparities in the usage patterns among the different social groups. By jointly using the unmanned aerial vehicles and ground photography to observe the pocket park usage patterns in Nanjing, we found that the senior and middle-aged groups are more inclined toward the activities like square dancing and chess compared to the youth and children who are keen on biking, running, and ball games [23]. Meanwhile, sampling surveys such as questionnaires and interviews provide a more comprehensive and in-depth understanding of group's specific usage habits and preferences. For example, Hussainzad [42] conducted a study on women's utilization tendency of green spaces in Kabul City of Afghanistan through questionnaires and found that the disparities of the social attributes, such as education, marital status, and employment status, significantly impact their use of green spaces. Accordingly, the clarification of the demand types of different social groups provides an important basis for determining service provision of pocket parks, as well as the construction of the service and spatial structure of pocket park groups.

5.3.2 Factors of Supply-Side

(1) Measurement of positive factors

According to Chapter 3, the positive factors in the built environment are aggregated by the recreation-related facilities, slow-traffic facilities, and partial types of public buildings. First, the recreation-related facilities can be categorized into public toilets, restaurants, and retail shops. The measurement of the recreation-related facilities includes proximity and density measurements [4, 6]. Usually, the areas with more diverse, closer, and denser recreation-related facilities in their surroundings show higher suitability for pocket park development. Second, the positive factors related to slow-traffic facilities mainly include slow-traffic lanes and bicycle parking lots [6, 43]. Similarly, the areas with closer and denser slow-traffic facilities in their surroundings are regarded as the more suitable sites for pocket park development. Third, pocket parks' service can also get support from surrounding public buildings like shopping malls, offices, and museums, which can offer abundant potential visitors and compensate for the functional deficiencies of pocket parks [4]. Accordingly, the development intensity and capacity of these buildings determine their support intensity to pocket park service.

With the increasing popularity of big data technology, the location and type information of most aforementioned facilities can be obtained through the open Application Programming Interface (API) of mapping platforms such as Baidu Maps, Gaode Maps, Tencent Maps, Google Maps, etc., which show higher precision and timeliness than government's statistical data [44]. For instance, in Shenzhen, Chen used the Baidu Map API to extract 11 types of facilities' points of interest (POIs), which were adopted to investigate the influencing factors on residents' visits of various parks [45]. Once the location and type information of the facilities are clarified, methods like kernel density estimation and proximity analysis can be applied to quantify their supportive effects on park service.

(2) Measurement of negative factors

For pocket park installation, one of the significant negative factors in the built environment is the proximate large-sized parks which show competition and inhibitory effect on pocket park service, as they will weaken the scarcity and uniqueness of the pocket parks to surrounding residents. The significance of this competitive effect is usually determined by parks' attractiveness to the residents of the same area [7]. Excluding parks' internal differences, the key determinants of this attractiveness are the size and proximity of the parks, as shown in the principles of the gravity model that are used for quantifying the accessibility [46]. It means that the potential areas of pocket park installation where the surrounding parks are larger or closer will have to face more intense competition from the surrounding parks. In addition, hazardous facilities in the built environment (such as distribution power stations and high-voltage lines) and intrusive facilities with strong noise or odor (such as garbage transfer stations) also negatively affect the services of pocket parks [47]. All these negative effects can be quantified by the models such the gravity model and its expanded version.

The integrated analysis and quantification of the positive and negative factors of the supply-side can provide a comprehensive assessment of the suitability of the pocket park development environment and lay the foundation for subsequent analysis on supply and demand fit.

5.3.3 Supply and Demand Fit

According to the theory of supply and demand of public resources [48, 49], the fit degree between the supply and demand sides is determined by the total level of the supply and demand intensity and their matching degree. The quantitative way to describe this feature in the site selection of pocket parks is that the areas whose values of the supply and demand conditions are both higher while the value difference between the two sides is lower will own higher fit degree. Otherwise, the areas show lower supply–demand fit degree. This also means that if a pocket park is installed at a site with higher supply–demand fit degree, it is not only helpful to promote the responsiveness to residents' demand, but also beneficial to get greater support from the surroundings in the built environment, by which the service performance of the pocket park can be guaranteed while the cost of pocket park construction and implementation can be saved.

5.4 Identification of Available Sites

The supply–demand fit analysis is able to identify the areas with high supply–demand fit degree, which are considered as the priority areas for pocket park installation. However, besides the supply–demand fit degree, factors like land acquisition opportunities and the construction and implementation conditions should also be considered in the decision-making on the site selection and service provision of pocket parks.

5.4.1 Land Acquisition Opportunities

Considering the basic conditions within built environment, pocket park development should prioritize the feasibility and cost-effectiveness and avoid large-scale demolition and construction. Therefore, in the areas with high supply–demand fit degree, the existing open spaces, such as the riverfront spaces, streetside spaces, abandoned lands, building entrance spaces, and interstice spaces between buildings (Fig. 5.2), should be firstly extracted as the priority alternative sites, as they show higher feasibility and lower cost for pocket park installation compared to other built-up lands or facilities. If the existing open spaces cannot meet the installation requirement, the green spaces attached to the public facilities or specific institutions (e.g., shopping malls, office buildings, universities, elementary and secondary schools, etc.) with semi-public attributes can also be considered as the alternative sites for developing all-day or time-sharing pocket parks (Fig. 5.3; Fig. 5.4).

If the lands acquired through the aforementioned methods remain insufficient for pocket park development, opening the internal green spaces or gardens of the gated communities to the public can be a viable alternative. For example, in the 1970s, Seattle developed a community gardening project "P-Patch" in the United States, aiming to open community gardens to the public as places for restoration, education, and gathering (Fig. 5.5) [50]. In practice, opening the access of green spaces within gated communities necessitates altering the management mode of the communities, which requires for the negotiation with the property owners and managers as well as the tailored operation and specific investment from the specialized agencies. For example, in 2010, Seattle government allocated an annual budget of US\$680,000 (about RMB 4.7 million) to 68 community gardens and set up a specialized agency taking charge of their management and operation [50]. However, the conflicts in the

(a) A streetside space (b) A riverfront space

Fig. 5.2 Existing underutilized spaces

Fig. 5.3 Middle school playground in Yicuiyuan community (Xi 'an, China) *Source* https://www.cnr.cn/sxpd/c/yl/20211025/t20211025_525642396.shtml

use of green spaces between local residents and outsiders are probably inevitable, requiring for additional coordination platforms or channels. This brings about more challenges in the implementation and management than the former two ways.

Moreover, sites with existing buildings and facilities that are to be relocated or renovated can also be seemed as alternative opportunities for pocket park installation. However, as their land acquisition process is more complicated and full of uncertainty, it asks for more cost in the relocation and remodeling work.

According to the feasibility of implementation and ease level of operation, the site selection for pocket park installation can be categorized into four levels in descending

Fig. 5.4 Middle school playground in Blanche Ford (Santa Cruz, USA) *Source* https://www.ssala.com/portfolio/branciforte-middle-school-sports-field-renovation

Fig. 5.5 P-patch in Greenwood (Seattle, USA) *Source* https://www.pccmarkets.com/sound-consumer/2020-05/planting-a-legacy-seattles-p-patch-gardens/

Table 5.1 Prioritization criterion for pocket park site selection

Prioritization	Source	Space form
First class	Existing open spaces	Waterfront spaces, streetside or street-center spaces, outdoor parking spaces, vacant or abandoned land, building entrance sites, and building leftover lots, etc
Second class	Attached green spaces to offices, schools, and other public institutions within	Playgrounds, courts, courtyards, internal squares, etc
Third class	Attached green spaces within neighborhoods	Central green spaces, community gardens, and other forms of community green spaces
Fourth class	Conversion from other land uses	Land with existing facilities or buildings

order: existing open space with low utilization rates, attached green space to public institutions, internal green spaces within community, and existing land conversion from other uses (Table 5.1).

5.4.2 Construction and Implementation

Besides development lands, pocket parks' service raises additional requirements on some specific attributes of the alternative sites, such as capacity, form, topography, etc.[4] Therefore, the site selection of pocket parks should also consider the site suitability to facilitate the construction work and service running. China's national *Standard for Planning of Urban Green Space* (2019) stipulates that the area of the mini-parks should not be less than 1000 m^2.[51] The *Public Open Space in Private Development Design and Management Guidelines* (2011) of Hong Kong raises specific requirements to the greenery area, aspect ratio, minimum width facing to the street, and size of the pocket spaces [52, 54]. According to the discussions in Chapters 2 and 3, we classified the influence of various factors regarding to site conditions into the types of positive, neutral, and negative (Table 5.2), which can provide references for pocket parks' site selection and service provision.

5.5 Grouping

The analyses of supply–demand fit and the identification of the alternative sites can provide alternative lands with high suitability for pocket park installation. However, determining the final sites and service mode of each pocket park should comprehensively involve the consideration of the park group construction within the service unit, rather than installing pocket parks separately. Therefore, the site selection and

Table 5.2 Influence of site conditions on pocket park installation

Site factors	Positive characteristics	Neutral characteristics	Negative characteristics	Affected aspects
Area	>4000m^2	400α~4000 m^2	<400m^2	Capacity, activity organization, and recreation experience
The open ratio of the perimeter	>75%	25%~75%	<25%	Openness, recreation experience
Shape aspect ratio	1:1~3:1	3:1~5:1	>5:1	Activity organization, recreation experience
Flatness of terrain	>80%	40%~80%	<40%	Convenience of space utilization, construction cost
Overall slope	<8%	8%~20%	>20%	Convenience of space utilization, construction cost, recreation experience, and activity organization
Orientation	South	East or west	North	Natural lighting, plant growing, recreation experience
Green coverage ratio	>65%	10%~65%	<10%	Construction cost, recreation experience
Other attachments	Facilities, buildings, or structures that can be used directly or through slight modification	Facilities, buildings, or structures that cannot be used but can be easily demolished	Hazardous, noisy, or odorous facilities that are difficult to be removed	Construction cost, use security, and recreation experience

service provision of the individual pocket park should also facilitate and promote the collaboration of the park groups. It is worth noting that although service collaboration is a type of mutual interaction among pocket parks within the group, parks' service objects and external environment also play key roles in shaping the collaboration due to the openness of the pocket park service system. Accordingly, the installation of pocket parks and construction of park groups are not only a matter of arranging the internal elements of the park groups but also related to the relationship rebalanced between parks and their service demands and external environment.

5.5.1 Demand Adaption

(1) Adaption to demand distribution

Since the pocket park is defined as a proximate server for outdoor recreation activities, its location relative to the location of service demand directly impacts its service efficiency. Similarly, the spatial interrelationship between the distribution of the pocket park group and the demand distribution within the service unit directly influences the spatial structure, collaboration modes, and demand response efficiency of the park group.

As proved in numerous studies and practices, the most effective mode of service provision of pocket parks is to locate them proximately to the demands [22, 40]. This can also be taken as the basic rule guiding the spatial distribution of pocket park group. Accordingly, in condition that the service demand is concentrated, the adjacent peripheral sites can be considered as the priority choices for pocket park installation, through which a ring—or radial-structured park group can be constructed to effectively deconcentrate the centrally-distributed demand. Contrarily, in condition that the service demand is scattered across the entire unit, it is more conducive to promote the group's efficiency by dispersedly locating each park to its own major service object, which commonly adopts the radial or network structure in space. Moreover, if the service demand distributes in linear mode, installing the pocket parks in linear structure to align with the demand distribution is always considered as the priority option (Fig. 5.6).

(2) Adaption to demand intensity

Due to the uneven distribution of the residents in space, the demand intensity varied significantly in different areas of the parks' service unit, raising differentiated requirements on the demand response of the pocket parks located in different places [40, 55]. Therefore, besides establishing an appropriate spatial structure according to the demand distribution, it is also crucial to configure a suitable size and capacity

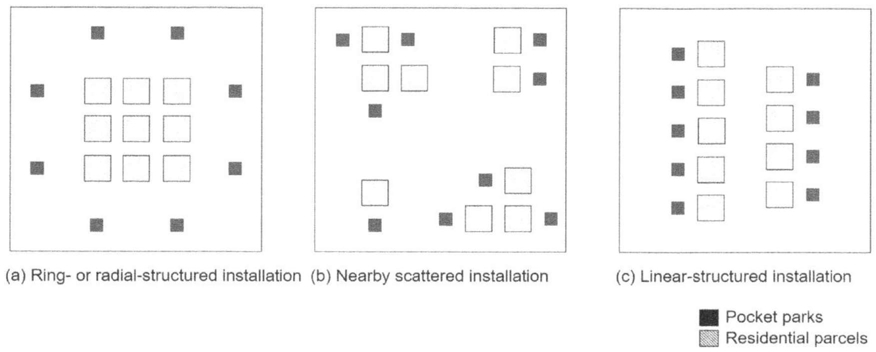

Fig. 5.6 Adaption modes of pocket park layouts to service demand distribution

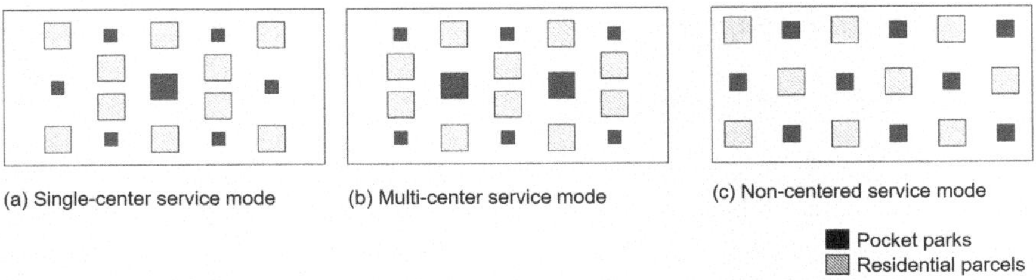

Fig. 5.7 Adaption modes of pocket park layouts to service demand intensity

for each pocket park in accordance with the demand intensity, by which a rational service structure can be constructed.

The construction of the service structure of pocket parks is primarily oriented by the distribution of the service demand intensity, which is mainly determined by resident density and group composition. For the service units where there are great differences in service demand intensity among different areas, parks with larger size and higher capacity should prioritize to be installed to the high-density areas, which easily forms a single-center or multi-center service structure. Contrarily, for the service units where the service demand intensity distributed evenly, the allocation of the park size and capacity can adopt a relatively balanced mode, commonly forming a non-centered service structure. (Fig. 5.7).

In the construction of the service structure, if pocket parks' size is constrained in space, increasing the proportion of the centralized venues can effectively expand the service capacity of pocket parks and mitigate the supply–demand imbalance. Additionally, installing multiple distributed parks surrounding the object service area where the service demand concentrates can also respond to the intense and centralized demand by a decentralized mode.

(3) Adaption to demand types

The variations in the composition of social groups result in significant differences in the types of service demand as well as the usage patterns of green spaces [40]. For instance, areas predominantly inhabited by seniors exhibit stronger demand for venues of centralized group activities and chess-card facilities in their surrounding space. Additionally, there is a stronger demand for constructing a slow-traffic ring to connect the pocket park group due to their specific demand on daily walking and jogging. Contrarily, areas primarily resided by youth and children demonstrate a greater demand for children's playgrounds and daily sports facilities in adjacent areas.

Therefore, the pocket parks with high fit degree between the types of service provision and service demands are more likely to be efficiently utilized. However, effective service collaboration does not equal to the high diversity of the spatial forms and facilities as well as the high differentiation of different parks' service. Instead, the recommended way is to formulate a customized proposal based on the comprehensive understanding of the demand types and their composition. For instance, in a service unit with high proportion of the seniors, the horizontal division mode in parks' service, such as evenly allocating parks' service capacity to various age groups, might fail to adequately fulfill the recreational demand of the seniors, who are supposed to be the majority park users of this unit. In such cases, using the composite service division mode, which shows sufficient flexibility in park use, richness in service patterns, and shareability for different groups, would be more effective in enhancing group's service efficiency. To achieve this goal, it is necessary to sophisticatedly investigate the travel time and demand pattern of each age group as well as the service capability of the parks, based on which a time-sharing service system can be established through the space and facility sharing among multiple groups simultaneously or in different periods. In this system, the seniors' demand can be prioritized to be responded while other groups' activity requirements can also be allowed. From the perspective of the service effectiveness and efficiency, the advantage of this approach is that the service of each park can be effectively integrated into the entire system, promoting their service flexibility and inclusiveness to different activities. Meanwhile, it also strengthens the internal connectivity of the park group in the service provision, thereby contributing to the enhancement of group's overall service efficiency (Fig. 5.8).

5.5.2 *Association with External Environment*

(1) Interaction with surrounding recreation-related facilities

Neighboring recreation-related facilities in the built environment can compensate for the service deficiency of pocket parks, by which pocket parks' service quality and integrity can be significantly improved [4, 6]. In the park group construction, installing pocket parks in association with the existing recreation-related facilities can promote the service efficiency of the park group and cut the construction cost, yielding twice the result with half the effort. However, in most cases, the recreation-related facilities are distributed unevenly in types and space. Thus, the service association between the pocket parks and their surrounding recreation-related facilities should be integrated into the site selection of each park according to its specific planning scenario. For example, the pocket parks mainly serving for the seniors or children should be preferentially installed in the areas with public toilets around. For pocket parks that mainly serve for middle-agers and youths, areas with surrounding cafes or restaurants should be given higher priority in site selection. For multi-service pocket parks that are available for group activities, priority should be given to the areas where commercial and retail facilities concentrated (Fig. 5.9).

On the other hand, the existence and aggregation of the specific facilities provide an important basis for the decision-making regarding to the service and spatial arrangement of the nearby pocket parks. For example, the pocket parks closed to the school entrances and exits could arrange a recreational and stopover site for waiting parents in children drop-off and pick-up, while those parks near shopping

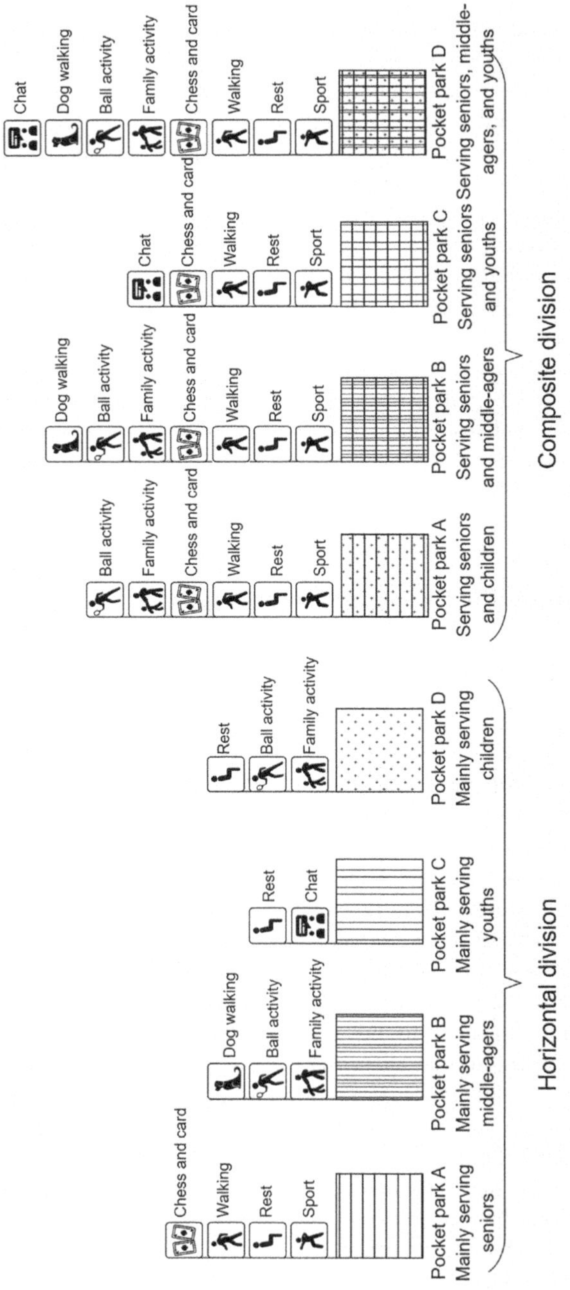

Fig. 5.8 Horizontal and composite division according to the demand types of different groups

(a) Pocket park for seniors: green space located northeast of Xiaowaxiang, Hongwu road (Nanjing, China)

(b) Pocket park for middle-agers and youths: green space near Congress Avenue (Texas, the USA) Figure source: http://colab.photo/congress-ave-pocket-parks

(c) Multi-service pocket park: green space near Jiangning Weaving Museum (Nanjing, China)

Fig. 5.9 Pocket parks distributed close to recreation-related facilities

malls and public facilities could specially set a centralized venue for group activities (e.g., exhibition, marketing, and performance). This service and spatial association with the surrounding facilities is beneficial to activating the interaction between parks and their surroundings. Meanwhile, the differences in the types and distribution of the existing facilities of different areas also provide good conditions for shaping the distinctive spaces within the parks and diversifying the services of the park group.

(2) Connection with existing slow-traffic networks

Existing slow-traffic network is not only the essential connections among pocket parks within a group, but also the important channels for visitors to access pocket parks. Therefore, the fabric of the service units and the structure of the slow-traffic network show essential influence on the site selection of pocket parks and the construction of the park group. For example, the service units with higher openness and denser slow-traffic lanes are seemed as the priority area for establishing the network structure. Contrarily, service units with lower openness and density of slow-traffic lanes are more suitable for adopting the linear or radial structure. In addition, existing studies also find that it is more conducive to park collaboration and visitor flow if more pocket parks can be distributed along residents' habitual walking routes, by which the frequency of park visit can be significantly promoted [56].

(3) Adaption with other types of open space

As mentioned in Chapter 3, complicated interactions exist between pocket parks and other types of public parks, green spaces, squares, and other outdoor open spaces in the built environment. Therefore, in condition that there are large open spaces within the service unit, such as urban parks (or comprehensive parks) and large squares, pocket parks should avoid being installed proximately to such spaces, by which the competition and inhibition effect on park service from these large open spaces can be alleviated. On the other hand, as an effective supplement to the conventional green spaces within high-density areas, pocket parks should also be integrated with the existing green spaces in function and space to form an efficient service system. For example, in the service units where the large open spaces are located, the site selection and grouping of pocket parks should differentiate themselves from the large open spaces in terms of the service provision and facility configuration. In most occasions, the pocket park group can serve as the sub-service nodes subordinating to the large open spaces, through which a multi-level recreational service network can be established (Fig. 5.10).

In addition to large open spaces, semi-open spaces (e.g., central green space of the gated neighborhoods) and large indoor recreational spaces (e.g., fitness centers, gymnasiums, etc.) also demonstrate substitutability of pocket parks [7, 57]. Our survey in the central city of Nanjing manifested that the service efficiency of the pocket park groups is negatively associated with the greening coverage of the service units [23]. Meanwhile, Peter Harnik also found that the indoor fitness facilities like gyms show competition effect on the utilization of nearby parks in US cities [7]. Therefore, to maximumly guarantee the service performance of the pocket parks, it is necessary to thoroughly consider the quantity, distribution, and service mode of other existing green spaces as well as the existing indoor recreational facilities (Fig. 5.11).

5.5.3 Coordination of Park Internal Constituents

(1) Service division

The service division of pocket park groups involves the works at the group and individual levels, which includes establishing the service division mode of the group and setting the service of the individual park, respectively. The establishment of the service division mode of park groups should not only respect the composition of social groups and their spatial distribution, but also consider the real conditions of the available spaces within the service units. For the service units where various social groups are highly mixed, vertical or horizontal division mode is preferred if there is sufficient development space while composite division mode is more suitable in condition that the development space is limited. For the service units where different social groups separately reside in different areas, the horizontal or composite division mode shows more suitability.

At the individual level, the dominant service setting of each pocket park should prioritize to respond to the service demand of the neighboring communities and maintain the consistency with the overall service division of the park group. However, the demand and service status of the park groups vary dynamically, which are commonly influenced by many unpredictable factors. For example, the pocket parks which are designated to serve for the seniors in peak hours also need to respond to the demands of various groups (such as passers-by, children, etc.) during off-peak hours [40]. Consequently, it is essential to strike a balance between the rigid and flexible settings in the service division of the park group.

The rigid setting can easily highlight the unique service characteristics of the individual pocket park, distinguishing it from other pocket parks. This is also the embodiment of the individuality of the pocket park, which is important to shape the

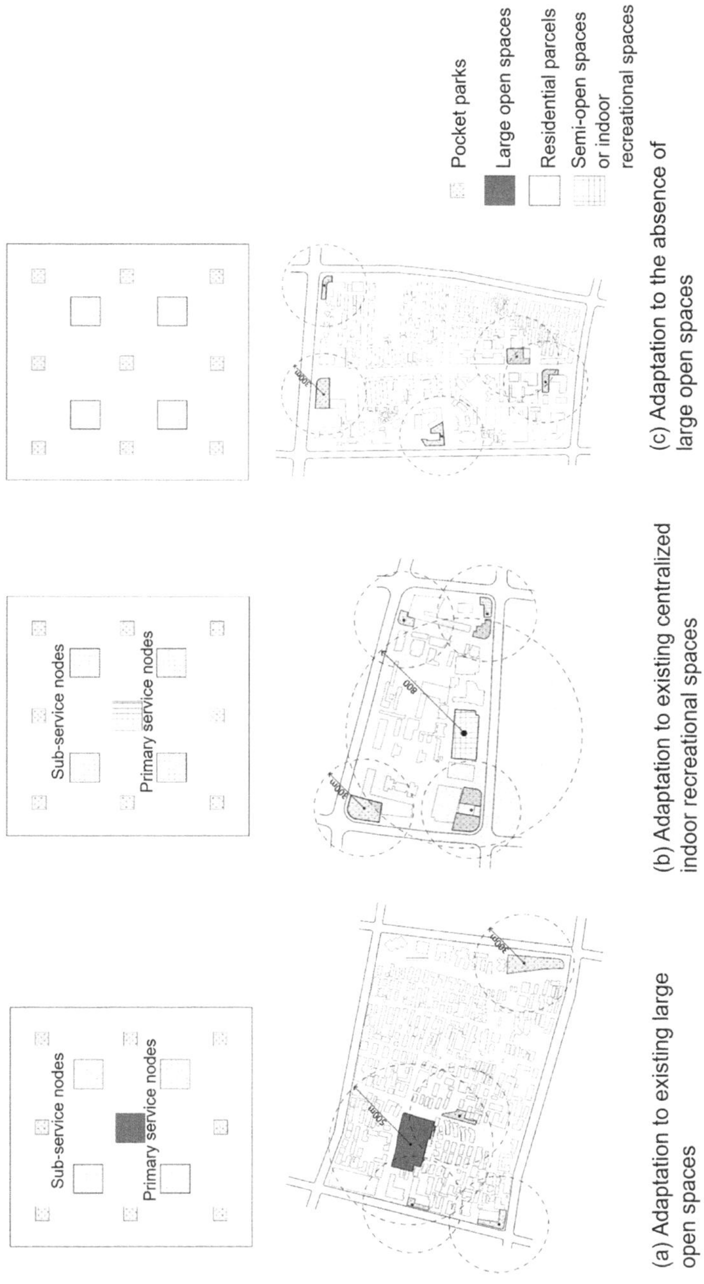

Fig. 5.10 Distributional adaption of pocket parks to other open spaces

(a) The public fitness center and its entrance

(b) The pocket park on the west of the Nanjing Federation of Trade Unions building

(c) The pocket park on the southwest of Jiangning Weaving Museum

(d) The pocket park on the southeast of Jiangning Weaving Museum

Fig. 5.11 [Nanjing, China] The public fitness center and surrounding pocket parks, southeast block of the crossing of Hongwu North Road and Changjiang Road

ability of the park group in serving diversely and complementarily. However, if the differentiation among pocket parks is too pronounced, it may result in a loss of service flexibility, ultimately limiting their service potential and performance. Contrarily, if some flexibility can be reserved for each park in the service division, it would be conducive for pocket park group to make moderate self-adjustment in their service process to adapt to the dynamic demand and other unpredictable factors, thereby reaching to a dynamic service balance by park collaboration.

(2) Sequence construction

The service sequence of pocket parks should be constructed along a continuous slow-traffic path, through which the sustained recreational experience can be provided for visitors [5]. Meanwhile, it is also important to make balance between the service division and correlation of the parks in the process of group construction [5] [Fig. 5.12].

The service division should highlight the dominant function of the individual park in the group while the correlation of the parks should focus on creating the commonalities and associations of the park group as a whole. The parks' dominant function can be jointly shaped by unique space organization and landscape element assembling. Meanwhile, to maintain the entirety and correlation of the parks in the service sequence, some commonalities or associations on the aspects of the forms of space organization, types of landscape elements, target social groups, and so on also need to be paid enough attention in park design and construction. For instance,

Fig. 5.12 [Nanjing, China] The proposed service sequence of Xiangpu Ying Community pocket park group and guidelines for the construction of the slow-traffic network

for the service sequence of pocket parks that mainly serves for middle-agers and seniors, differentiated spatial organization and facility configuration can be done in each park, by which different parks can provide various services in peak hours, such as chess and card activities, individual dance and exercise, group dance and exercise, and pet walking, providing diverse choices for the target groups [58] [Fig. 5.13]. Meanwhile, these parks can also be connected by the same slow-traffic lane, along which the forms and materials of the street furniture, amenities, and signages can be designed in a unified style, highlighting the coherence, commonalities, and entirety of the parks in the sequence (Fig. 5.13).

(3) Complementarity of facilities

Although pocket parks may get the service assistance from the external recreation-related facilities, it does not mean the external facilities can completely replace parks' internal facilities. If the site condition allows, more internal facilities installed within pocket parks can bring greater convenience to park users and enhance the quality of park service [38].

Due to the constraints in space, it is difficult for a single pocket park to completely configure all the types of service facilities as the urban parks do. Accordingly, to achieve an effective collaboration of park group, it is recommended to distribute the specialized service facilities into different parks according to the dominant service of each park. For example, pocket parks that mainly serve for the seniors can prioritize the installation of toilets, benches, and fitness facilities; pocket parks that mainly serve for children can prioritize the installation of playing and amusement facilities; pocket

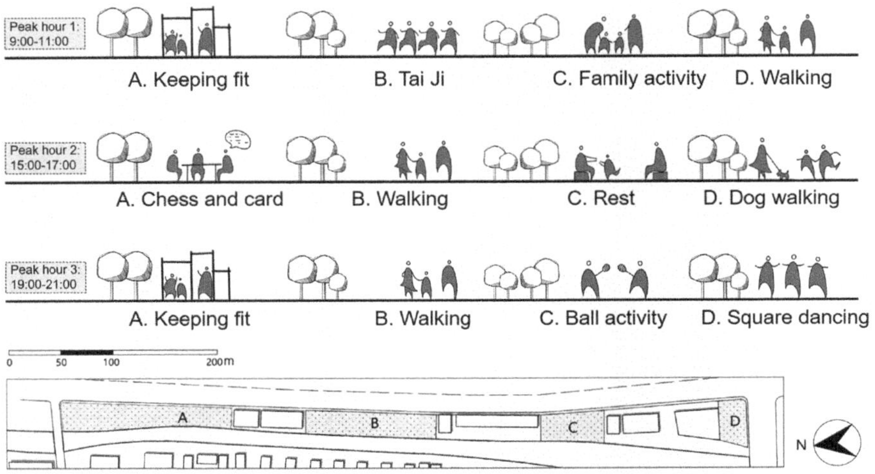

Fig. 5.13 Construction model of a composite service sequence for pocket park groups

parks that mainly serve the middle-agers and youth can prioritize the installation of facilities for stopover and social communication, such as cafes.

From the perspective of park group construction, this facility configuration mode is beneficial to form a completed facility service system of the whole group, thereby compensating for the individual imperfect of facility types by the complete facility system in group level. This mode can also promote the complementarity among pocket parks within a group and the flow of users, further strengthening the collaboration effect of the park group (Fig. 5.14; Fig. 5.15).

Fig. 5.14 Similarities in the forms of slow-traffic paths, seats and pavements, along the Yangtze River Road (Nanjing, China)

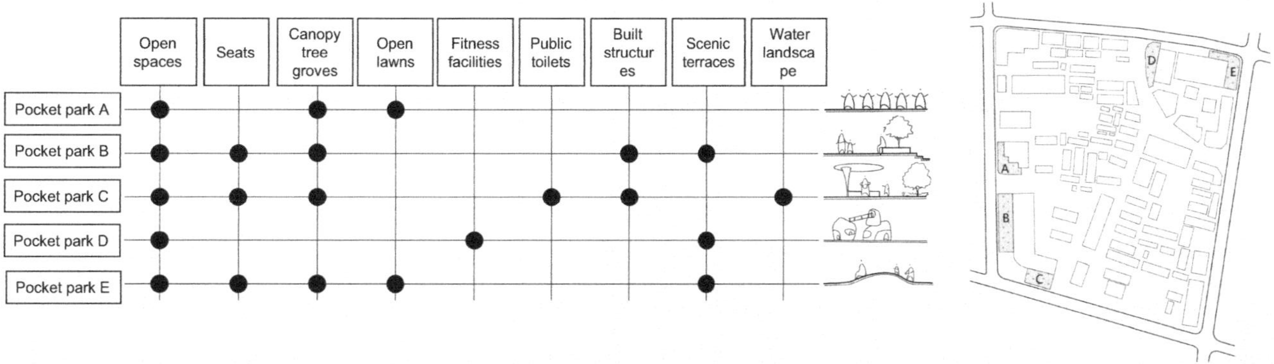

Fig. 5.15 The coordination arrangement of recreational service facilities in pocket park groups

References

1. Derdouri A, Murayama Y, Morimoto T et al (2025) Urban green space in transition: A cross-continental perspective from eight Global North and South cities. Landsc Urban Plan 253:105220
2. Liu S, Wang X (2021) Reexamine the value of urban pocket parks under the impact of COVID-19. Urban Forestry & Urban Greening 64:127294. https://doi.org/10.1016/j.ufug.2021.127294
3. Zhang S, Zhou W (2018) Recreational visits to urban parks and factors affecting park visits: Evidence from geotagged social media data. Landsc Urban Plan 180:27–35. https://doi.org/10.1016/j.landurbplan.2018.08.004
4. Forsyth, A., & Musacchio, L. (2005). *Designing small parks: A manual for addressing social and ecological concerns*. John Wiley & Sons.
5. Rioux L, Werner CM, Mokounkolo R et al (2016) Walking in two French neighborhoods: A study of how park numbers and locations relate to everyday walking. J Environ Psychol 48:169–184. https://doi.org/10.1016/j.jenvp.2016.10.003
6. Gehl, J. (2001). *Life between buildings: Using public space*. The Danish Architectural Press.
7. Harnik P, Green U (2010) Innovative parks for resurgent cities. Island Press
8. Boulton C, Dedekorkut-Howes A, Byrne J (2018) Factors shaping urban greenspace provision: A systematic review of the literature. Landsc Urban Plan 178:82–101. https://doi.org/10.1016/j.landurbplan.2018.05.029
9. Wang D, Brown G, Liu Y (2015) The physical and non-physical factors that influence perceived access to urban parks. Landsc Urban Plan 133:53–66. https://doi.org/10.1016/j.landurbplan.2014.09.007
10. Rawls J (1971) A theory of justice. Cambridge University Press
11. Wolch JR, Byrne J, Newell JP (2014) Urban green space, public health, and environmental justice: The challenge of making cities 'just green enough.' Landsc Urban Plan 125:234–244. https://doi.org/10.1016/j.landurbplan.2014.01.017
12. Da Silva RGP, Lima CL, Saito CH (2023) Urban green spaces and social vulnerability in Brazilian metropolitan regions: Towards environmental justice. Land Use Policy 129:106638. https://doi.org/10.1016/j.landusepol.2023.106638
13. Kabisch N, Haase D (2014) Green justice or just green? Provision of urban green spaces in Berlin, Germany. Landsc Urban Plan 122:129–139. https://doi.org/10.1016/j.landurbplan.2013.11.016
14. West ST, Shores KA, Mudd LM (2012) Association of available parkland, physical activity, and overweight in America's largest cities. J Public Health Manag Pract 18(5):423–430
15. Li F, Yao N, Liu D et al (2021) Explore the recreational service of large urban parks and its influential factors in city clusters–Experiments from 11 cities in the Beijing-Tianjin-Hebei region. J Clean Prod 314:128261
16. Zhang L, Guo C, Jiang Y (2018) Research on the key problems of leisure service radius of urban green space. IOP Conference Series: Earth and Environmental Science 170:032005
17. Chen Y, Ge Y, Yang G et al (2022) Inequalities of urban green space area and ecosystem services along urban center-edge gradients. Landsc Urban Plan 217:104266
18. Ke X, Huang D, Zhou T et al (2023) Contribution of non-park green space to the equity of urban green space accessibility. Ecol Ind 146:109855
19. Węziak-Białowolska D (2016) Quality of life in cities–Empirical evidence in a comparative European perspective. Cities 58:87–96
20. Schilling, J. (2010). Towards a greener green space planning. *International Master's Programme in Environmental Studies and Sustainability Science*.
21. Nordh H, Hartig T, Hagerhall CM et al (2009) Components of small urban parks that predict the possibility for restoration. Urban Forestry & Urban Greening 8(4):225–235. https://doi.org/10.1016/j.ufug.2009.06.003
22. Akpinar A (2016) Factors influencing the use of urban greenways: A case study of Aydın, Turkey. Urban Forestry & Urban Greening 16:123–131. https://doi.org/10.1016/j.ufug.2016.02.004

23. Zhou, C., An, Y., Zhao, J., et al. (2022). How do mini-parks serve in groups? A visit analysis of mini-park groups in the neighbourhoods of Nanjing. *Cities*, null. https://doi.org/10.1016/j.cities.2022.103804
24. Huang J, Wang X, Zhang D et al (2023) Trade-offs under pressure? Development of urban green space under economic growth and governance. J Clean Prod 427:139261
25. Yang J, Alahmadi KM, Shirowzhan S (2022) The impact of urban expansion in Beijing and Metropolitan Area urban heat Island from 1999 to 2019. Applied Mathematics and Nonlinear Sciences 7(2):581–600
26. Liu, W., Yang, J., Gong, Y., et al. (2022). An evaluation of urban renewal based on inclusive development theory: The case of Wuhan, China. *ISPRS International Journal of Geo-Information, 11*(11), 563.
27. Cui, D., Ditta, A. A., & Cao, S. J. (2024). Energy justice and sustainable urban renewal: A systematic review of low-income old town communities. *Journal of Cleaner Production*, 143470.
28. Gao, J., Qiao, W., Ji, Q., et al. (2021). Intensive-use-oriented identification and optimization of industrial land readjustment during transformation and development: A case study of Huai'an, China. *Habitat International, 118*, 102451.
29. Denhardt RB, Denhardt JV (2000) The new public service: Serving rather than steering. Public Adm Rev 60(6):549–559. https://doi.org/10.1111/0033-3352.00117
30. Denhardt RB, Denhardt JV (2001) The new public service: Putting democracy first. Natl Civ Rev 90:391
31. Boudreau JA (2022) City of repair: Practicing the future in Mexico City. Int J Urban Reg Res 46(5):851–864
32. Arbaci S, Tapada-Berteli T (2012) Social inequality and urban regeneration in Barcelona city centre: Reconsidering success. European Urban and Regional Studies 19(3):287–311
33. Liu J, Li G (2021) Comprehensive benefit evaluation on urban village micro-transformation based on extension matter-element model. Int J Constr Manag 21(5):524–537
34. Cohen DA, Marsh T, Williamson S et al (2010) Parks and physical activity: Why are some parks used more than others? Prev Med 50:S9–S12. https://doi.org/10.1016/j.ypmed.2009.08.020
35. Feltynowski M, Kronenberg J, Bergier T et al (2018) Challenges of urban green space management in the face of using inadequate data. Urban Forestry & Urban Greening 31:56–66
36. Cohen DA, Han B, DeRose KP et al (2017) Promoting physical activity in high-poverty neighborhood parks: A cluster randomized controlled trial. Soc Sci Med 176:150–157
37. Han H, Lai SK (2012) National land use management in China: An analytical framework. Journal of Urban Management 1(1):3–38
38. Ibes DC (2015) A multi-dimensional classification and equity analysis of an urban park system: A novel methodology and case study application. Landsc Urban Plan 137:122–137. https://doi.org/10.1016/j.landurbplan.2014.12.014
39. Zhou C, Zhang Y, Fu L et al (2021) Assessing mini-park installation priority for regreening planning in densely populated cities. Sustain Cities Soc 67:102716. https://doi.org/10.1016/j.scs.2021.102716
40. Cohen DA, Han B, Nagel CJ et al (2016) The First National Study of Neighborhood Parks: Implications for physical activity. Am J Prev Med 51(4):419–426. https://doi.org/10.1016/j.amepre.2016.03.021
41. Danford RS, Strohbach MW, Warren PS et al (2018) Active greening or rewilding the city: How does the intention behind small pockets of urban green affect use? Urban Forestry & Urban Greening 29:377–383. https://doi.org/10.1016/j.ufug.2017.11.014
42. Hussainzad EA, Yusof MJM, Maruthaveeran S (2021) Identifying women's preferred activities and elements of private green spaces in informal settlements of Kabul city. Urban Forestry & Urban Greening 59:127011. https://doi.org/10.1016/j.ufug.2021.127011
43. Kang Y, Fukahori K, Kubota Y (2018) Evaluation of the influence of roadside non-walking spaces on the pedestrian environment of a Japanese urban street. Sustain Cities Soc 43:21–31. https://doi.org/10.1016/j.scs.2018.07.009

44. Liu Y, Liu Y (2022) Detecting the city-scale spatial pattern of the urban informal sector by using the street view images: A street vendor massive investigation case. Cities 131:103959. https://doi.org/10.1016/j.cities.2022.103959
45. Chen Y, Liu X, Gao W et al (2018) Emerging social media data on measuring urban park use. Urban Forestry & Urban Greening 31:130–141. https://doi.org/10.1016/j.ufug.2018.02.005
46. Neutens T, Schwanen T, Witlox F et al (2010) Equity of urban service delivery: A comparison of different accessibility measures. Environment and Planning A: Economy and Space 42(7):1613–1635. https://doi.org/10.1068/a4230
47. Kerishnan PB, Maruthaveeran S (2021) Factors contributing to the usage of pocket parks-A review of the evidence. Urban Forestry & Urban Greening 58:126985. https://doi.org/10.1016/j.ufug.2021.126985
48. Bartesaghi-Koc C, Osmond P, Peters A (2019) Spatio-temporal patterns in green infrastructure as driver of land surface temperature variability: The case of Sydney. Int J Appl Earth Obs Geoinf 83:101903. https://doi.org/10.1016/j.jag.2019.101903
49. Rupprecht CD, Byrne JA, Ueda H et al (2015) 'It's real, not fake like a park': Residents' perception and use of informal urban green-space in Brisbane, Australia and Sapporo, Japan. Landsc Urban Plan 143:205–218. https://doi.org/10.1016/j.landurbplan.2015.07.002
50. Glennie C (2020) Cultivating place: Urban development and the institutionalization of Seattle's P-Patch community gardens. City Community 19(3):726–746. https://doi.org/10.1111/cico.12378
51. Ministry of Housing and Urban-Rural Development of the People's Republic of China. (2019). Standard for planning of urban green space: GB/T51346–2019 [S]. China Architecture & Building Press.
52. Cooper Marcus, C., & Francis, C. (1998). People places: Design guidelines for urban open space. Wiley.
53. Development Bureau of the Government of the Hong Kong. (2011). Public Open Space in Private Developments Design and Management Guidelines [S]. Hongkong.
54. Seymour Jr, W. N. (1969). Small urban spaces: The philosophy, design, sociology, and politics of vest-pocket parks and other small urban open spaces.
55. Hamstead ZA, Fisher D, Ilieva RT et al (2018) Geolocated social media as a rapid indicator of park visitation and equitable park access. Comput Environ Urban Syst 72:38–50. https://doi.org/10.1016/j.compenvurbsys.2018.01.007
56. Harper, J. (2009). Planning for recreation and parks facilities: Predesign process, principles, and strategies. Venture Publishing, Inc. https://search.library.wisc.edu/catalog/9910467099002121
57. Armato F (2017) Pocket Park: Product urban design. Des J 20(sup1):S1869–S1878. https://doi.org/10.1080/14606925.2017.1352705
58. Nordh H, Østby K (2013) Pocket parks for people—A study of park design and use. Urban Forestry & Urban Greening 12(1):12–17. https://doi.org/10.1016/j.ufug.2012.11.003

Chapter 6
Institutional Guarantee

6.1 Introduction

Apart from an effective planning and design framework, every successful installation of pocket park is the result of numerous rounds of coordination and negotiation, which includes the coordination within the planning system [1], cooperation with the legislative and administrative system [2], and negotiation with the stakeholders [3]. However, we cannot expect these activities to spontaneously occur and proceed in an orderly way. Thus, to smoothly facilitate these activities and guarantee their effectiveness, it is necessary to establish a well-structured institutional guarantee system. As the countries with different legislative and administrative systems may differ in the construction and implementation of the institutional guarantee system, this chapter initiates the discussion mainly based on the current system of China in order to clearly demonstrate the occurrence background of most case studies of this book. Moreover, the successful system design, implemental approaches, and cases of other countries are also introduced to enlighten the relevant practices and improvements of current systems.

6.2 Planning System

According to the current framework in the territorial and spatial planning of China, the control of pocket park installation can be divided into four aspects: quantity control, group structuring, individual positioning, and boundary control. The relevant planning approaches involve three levels of planning: master planning, sector planning, and detailed planning, each of which has its own highlights in the control of pocket park installation.

C. Zhou, *Principles and Methods of Pocket Park Planning and Design*, Urban Sustainability, https://doi.org/10.1007/978-981-96-5801-5_6

6.2.1 Master Planning

Master planning focuses on the comprehensive deployment and implementation of measures for city functions, development goals, city size, land use, spatial layout, and various constructions within a preset period [4]. It is grounded in macro-level considerations, encompassing the comprehensive integration and general arrangement of the total quantity and spatial structure of various categories of land use, including parklands. Due to the tiny size and non-independence in land use for partial pocket parks, it is difficult to directly reserve the lands for all the pocket parks in the level of master planning. The planning approaches related to pocket parks mainly focus on the control of the overall quantity and individual location of the small-sized parkland parcels, which can be titled as pocket parks of "formal green spaces" (Table 6.1).

On the other hand, the factors and elements related to pocket parks' site selection and service provision are preliminarily formed in master planning, such as the layout of residential areas that generally determines the spatial distribution of the service objects of pocket parks, the road system that totally divides the service units of pocket park groups, and the public service facility system that provides the various external support for pocket park service. These also establish an essential basis for the formulation of the planning proposal of pocket parks in the subsequent planning process.

6.2.2 Sector Planning

Sector planning is a plan prepared for a specific sector of national economic and social development, which involves the unfolding, deepening, and concretization of some major aspects and key areas outlined in the master planning [4]. Urban green space system planning is a sector planning that touches the site selection and service provision of pocket parks.

In urban green space system planning, the function, indicators, structure, and site of each type of urban green spaces are systematically arranged [5]. Besides parklands that own independent land parcels, it also sets the planning indicators for the green spaces attached to other lands and formulates the guidelines to control their spatial structure. Therefore, this planning can comprehensively coordinate the quantity and structure of all elements related to pocket parks. Accordingly, following the control framework of master planning, this planning not only further fixes the size and locations of the pocket parks under the category of "formal green space", but also controls the total quantity of pocket parks under the category of "informal green space".

In recent years, with a growing emphasis on the city's sustainable development and residents' green travel, many cities have further extended the sector planning to the field of urban greenway planning, public park planning, and so on, building upon the urban green space system planning [6, 7]. Urban greenway planning can make

Table 6.1 Key points in various plannings related to pocket park installation

Planning	Type	Pocket parks of "formal green space"				Pocket parks of "informal green space"			
		Quantity control	Group structure	Individual location	Boundary control	Quantity control	Group structure	Individual location	Boundary control
Master planning	Urban master planning	●	–	○	–	–	–	–	–
Sector planning	Urban green space system planning	▲	●	●	○	●	●	○	○
Detailed planning	Regulatory planning	▲	●	▲	●	▲	●	△	△
	Site planning	▲	▲	▲	▲	▲	▲	●	●

Note ●means that the content is controlled in the planning; ○means that the content is guided in the planning; ▲means that the content of the representation is refined and controlled by the upper planning; △means that the content follows the upper planning for detailed guidance; —means that the content is not mentioned in the planning corresponding arrangements to construct the spatial links among pocket parks based on the slow-traffic network, while the pocket parks along the greenways can be regulated and controlled as the service nodes of the entire system. Meanwhile, other sector plannings like public park planning can contribute to the more comprehensive and in-depth arrangements for the urban outdoor recreational system, which not only makes an in-depth control and guidance over the quantity, layout, and pattern of pocket parks, but also establishes some more specific functional and layout coordination between pocket parks and other public parks.

6.2.3 Detailed Planning

Detailed planning involves the exact arrangements of land use, spatial form, and construction in a specified period of the local areas, which includes two phases: regulatory planning and site planning [8, 9].

(1) Regulatory planning

Regulatory planning, almost equivalent with the US zoning, primarily follows the relevant stipulations of master planning to determine the land use category and intensity, placements of roads and infrastructure pipelines, and control requirements of the environment [10]. Regulatory planning is critical to implementing the spatial structure and service division of pocket park groups raised in sector planning. On one hand, it accurately controls the location and boundary of pocket parks under the category of "formal green space" and guide their design style and details. On the other hand, it also provides guidelines for controlling the location, boundary, and construction features of pocket parks under the category of "informal green space". Additionally, regulatory planning encompasses the control of land development intensity (which directly determines the intensity of service demand of pocket parks), types and location of recreational-related facilities, and layout of park links.

(2) Site planning

The main task of site planning is to propose detailed arrangements and designs for the site construction based on the planning conditions offered by the regulatory planning [10]. For the site selection and service provision of pocket parks, this planning can provide clear control and detailed arrangements for the location and boundaries of the "informal" pocket parks, and thoroughly implement the design and construction guidelines outlined in the regulatory planning.

6.3 Control Approaches

6.3.1 Regulations and Indicators

(1) General regulations and indicators

The control regulations or planning standards promulgated by the competent authorities provide direct guarantees for the planning and implementation of pocket parks. In the early urban green space planning in China, there is merely baseline land use quota to guarantee the parkland areas, such as the minimum per capita area of parkland (Table 6.2). However, due to the absence of the subdivided indicators to control the size of parks in different types or classes, it can be frequently found an imbalanced distribution of park types in the cities. For example, in most early urbanization process of China, it can be easily observed that resources are overly used in the development of the large-sized parks at the higher class while ignoring a balanced development of small-sized green spaces at the lower class[11]. This condition did not change until the 2010s, when the central government begun to pay more attention to the urban development quality and advocate the organic city renewal practice [12, 13].

To improve the green space service structure and facilitate a balanced development of various green space types in the urban renewal, the national *Standard for Planning of Urban Green Space (GB/T 51346–2019)* was enacted [14]. This standard firstly provides subdivision indicators of each park type and class, which requires that the per capita area of mini-parks, one type of pocket parks under the category of formal green spaces, should be no less than 1 m^2. This is also the first time that the baseline size of pocket parks is guaranteed in the citywide scale. At the community level, the *Standard for Urban Residential Area Planning and Design (GB 50180–2018)* has also made corresponding regulations regarding to the individual size (0.4 ~ 1 hm^2) and per capita area (not less than 1 m^2) of small-sized green spaces within the scope of "5-min pedestrian-scale neighborhoods" [15].

In addition to ensuring and controlling the park area, the guarantee of the rationality of the spatial distribution of pocket parks can be done through the baseline requirements or indicators of accessibility. For example, *Guidelines on promoting the work of Urban Renovation and Ecological Restoration of the Ministry of Housing and Urban–Rural Development* [16], released in 2017, and *the Series Standards of National Garden City* [17], enacted in 2016, listed the baseline accessibility requirements and indicators, respectively, as mentioned in Subsection 5.4.2. Once the accessibility indicator is incorporated into the planning as a mandatory requirement, it can not only guarantee the basic quantity of pocket parks, but also lead to a balanced result in parks' spatial distribution, particularly contributing to improving the service coverage and quality of green spaces in the areas where the existing green resource is scarce.

(2) Specified regulations and indicators

Table 6.2 Patterns and methods of pocket park development guarantee

Control modes	Methods	Control focuses	Representative indicators or requirements	Example
General control	National regulations and standards	Quantity	The minimum indicator of per capita area of mini-parks, the lowest are proportion of mini-parks in the total area of parklands	Standard for Planning of Urban Green Space (GB/T 1346–2019)
		Accessibility	Service radius indicators and specific accessibility requirements	Guidelines on promoting the work of Urban Renovation and Ecological Restoration of the Ministry of Housing and Urban–Rural Development (2017)
		Quality	Facility configuration regulations and greening coverage indicators	Code for Design of Public Park (GB 51192–2016)
		Compensatory mechanism	Compensatory standards of small-sized green spaces for the absence of large parks	Standard for Urban Residential Area Planning and Design (GB 50180–2018)
Specified control	Local regulations, standards, and specific project control	Development opportunity	Utilization of the setback spaces of streetside buildings and Utilization of the setback spaces in the center of each plot in non-independently occupied blocks	Public Open Space System Planning of Shenzhen Special Economic Zone (2007)
		Quantity	Ratio of area between small public spaces and other functional spaces and minimum size of individual park	Downtown Plan in San Francisco, USA (1984)
		Design	Parks' orientation, form, vegetation, facilities, and style	Public Open Space in Private Developments Design and Management Guidelines in Hong Kong (2011)

Applying more specified control measures or indicators at the project level can further guarantee the development opportunities of pocket parks in high-density areas as well as the rationality of parks' quantity and location arrangements. For example, the *Downtown Plan* of San Francisco, formulated in 1985, stipulated that the provision of every 50 square feet (about 4.6 square meters) of new office space should ensure the provision of 1 square foot (about 0.093 square meter) of accessible open space for public use [18]. In China, cities like Shenzhen, Hangzhou, and Wuhan stipulated in their sector plannings for public spaces or regulatory plannings that, in the non-independently occupied blocks, the building boundaries of each land plot (e.g., whose land area is larger than 8,000 square meters) should be set back to form a centralized open space (e.g. not less than 400 square meters) [19] [Fig. 6.1]. Moreover, cities such as Shanghai, Guangzhou, and Foshan of China listed specified control indicators and strategies in their design guidelines for streets or open spaces to encourage the use of setback spaces of the streetside buildings in micro green space development [20] (Fig. 6.2). The existence and use of specified regulations and indicators reserve necessary flexibility for making the tailored control of different local governments, as the density levels and realistic conditions of different cities greatly differ.

In addition to the aforementioned approaches, to achieve the goals of the collaborative service for pocket park groups, specified regulations and indicators also need to be introduced in terms of the associated distribution, interconnection, and service sequence construction of pocket park groups. As this control and guidance demonstrate relatively high requirements on the precision and environmental adaptability, it also asks for a meticulous examination of the characteristics of the planned area.

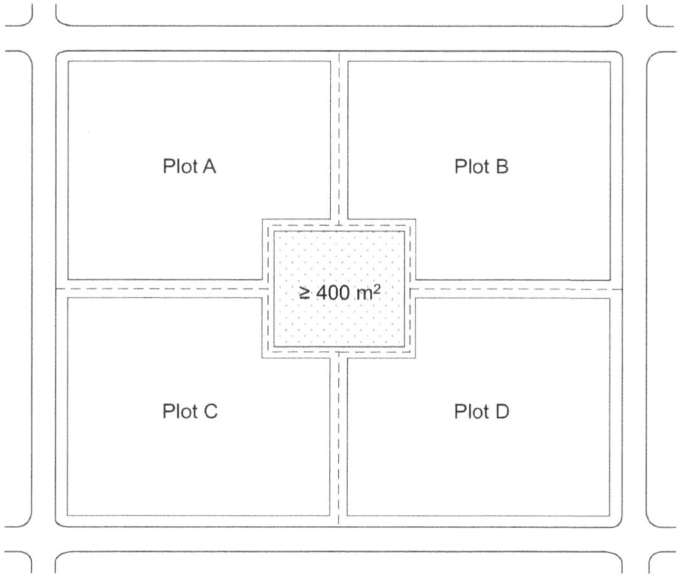

Fig. 6.1 Shared central open space generated by joint setback of surrounding plots

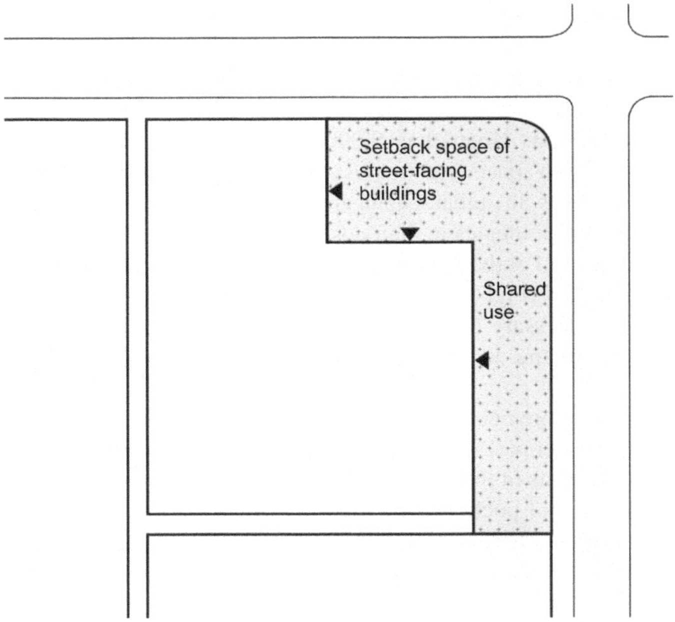

Fig. 6.2 Shared corner open space generated by the setback of street-facing buildings

6.3.2 Cross-Class Compensation Mechanisms

Apart from the rigid control regulations and indicators, establishing a mechanism with greater flexibility for the mutual compensation of the green spaces in different classes can also promote the development of pocket parks, especially in high-density areas. This is mainly because that the emphasis on the hierarchical system of green space services or rigid control indicators deviated from the realistic conditions of the high-density areas. For example, in most areas of the old cities or historic districts with high building density, it is less feasible to meet the requirements regarding to some green space control indicators, such as the stipulation on the per capita area of the community park and their individual size.

This also leads to a fact that the conventional hierarchical service system will lose efficacy or cannot run as expected in the high-density environment where the flattened characteristics in the service of the green spaces of different classes or sizes are strengthened [21]. This easily results in the horizontal mode of service division among parks, creating good conditions for the mutual compensation among the green spaces with different classes. For example, the pocket park system proposed by Robert Zion in the 1960s, as mentioned in Chapter 1, was originally designed for compensating for the absence of large-sized parks in the city center [22]. The essence of this approach is to use a flattened and distributed service system composed by multiple small-sized green spaces to replace a large park that serves in centralized mode. China's *Standard for Urban Residential Areas Planning and Design (GB/T50180-2018)* [15] also applies this principle and allows to use multiple small-sized green spaces to substitute the larger community parks once the land use condition cannot allow for the development of larger parks in the city renewal.

Similarly, this compensation can also run in a reverse way. William Whyte discovered that the large-sized parks such as the Central Park of New York could be split into a series of small spaces in its recreational service process, which is also consistent with the way that people perceive the spaces within the park [23]. Accordingly, in the high-density cities, a large-sized green space can also be regarded as an aggregation composed by multiple small-sized green spaces or pocket parks, and thus demonstrate the potential to form a daily service sequence once the small-sized green spaces are insufficient or absent. It is notable that the compensation of the large-sized green spaces for the small-sized green spaces is to achieve the goal of the intensive and efficient utilization of the green resources, rather than constrain the development of the latter. Meanwhile, if the service space (or node) sequence within the large-sized green space can be finely decomposed, reclassified, and redefined, making them differentiated from the green spaces around, it can be expected that all the recreational service spaces within and around the large-sized green space jointly construct a well-organized and mutually complementary service sequence for the entire district, thereby averting the competition or inhibition effect among green spaces of different sizes (Fig. 6.3).

However, due to the complex interaction mode among green spaces, there is still abundant work to be done to improve the effectiveness and precision of current compensation mechanism among green spaces of different sizes. Particularly, more future explorations need to be focus on revealing the interaction patterns between different green spaces and the generation mechanism of the compensation effects, so as to provide an objective basis for setting the compensation indicators.

6.3.3 Integral Control of Informal Green Space

Most statistical data and planning indicators of urban green spaces merely centered on formal green spaces, while informal green spaces, mainly composed by attached green spaces, are difficult to be measured by current planning indicators and cannot be involved in the statistical data, generating great challenges to planning control and balance. However, both the number and area of the attached green spaces usually accounts for more than half of the urban green spaces [24], for which the pocket parks from attached green spaces also play an irreplaceable role in the daily recreation services. This directly leads to a serious inconsistency between the results oriented by the in-used planning indicators and the actual service effect of all the green spaces.

Accordingly, it is important to set specific indicators to control the quantity and service status of the attached green spaces (or informal green spaces) that can serve for the public. Moreover, these indicators should also be convenient to be integrated with the original measurement system of formal green spaces, by which the pocket

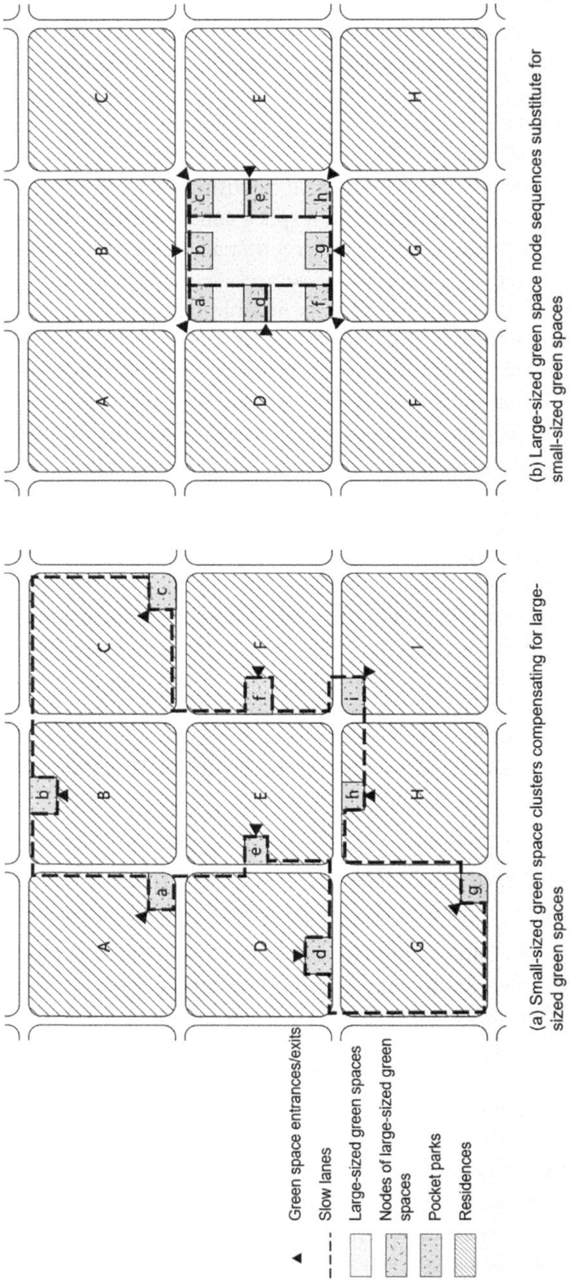

Fig. 6.3 Cross-size compensation mechanism of green spaces

parks from the informal green spaces and other elements that have potential to be transformed into such green spaces can be involved into the decision-making system of urban planning (Table 6.3). For example, the San Francisco Planning + Urban Research Association (SPUR) has established a detailed database of Privately Owned Public Open Spaces (POPOS), which serves as an important basis for the future optimization of city's park system in relevant plannings [25]. Moreover, the service radius of pocket parks under the category of informal green space is seemed as another essential indicator in planning, which is usually set as 5-min walking distance (200 ~ 400 m) in European and North American cities [26, 27]. This setting will also facilitate an integrate distribution of the pocket parks under both categories of the formal and informal green spaces in planning.

Once the planning indicators and statistical mechanism for informal green spaces are set up, all the available resources can be effectively mobilized and integrated in the process of the spatial distribution and service division of green spaces in planning. This approach can also effectively prevent the repeated construction of green spaces and the wastage of resources and strengthen the consistency between the indicative information of the planning indicators and the actual effect of green space services.

6.4 Support Policies

The government can also support the development of pocket parks by formulating targeted support policies, which construct the base for the rapid development of pocket parks [28]. From the perspective of policies' action modes, they can be categorized into the policies of direct support and indirect encouragement. Direct support policies usually involve the government budgeting, which can be directly used in the land acquisition, construction, and maintenance regarding to pocket parks [29]. Conversely, indirect approaches focus on the formulation of the encouragement policies such as development compensation and opening encouragement to motivate the other institutions or private organizations to develop pocket parks [30].

6.4.1 Direct Support

The government's direct support policies commonly allocate more support to the development of pocket parks under the title of formal green spaces which have their own land parcels, because these parcels can be easily supervised and handled by governments. Since the development of pocket parks in built-up areas mainly relies on the existing lands, the acquisition of development land becomes a prerequisite for the implementation of planning proposal. For western cities whose lands are privately owned, the inevitable issues in the development and installation of pocket parks are the transition of land ownerships or the lease of land-use rights. Accordingly,

Table 6.3 Statistical table of urban green space planning incorporating "informal green spaces"

Codes	Types		Quantity		Area		Per capita area		Percentage of urban built-up land	
			Status quo	Plan	Status quo	Plan	Status quo	Plan	Status quo	Plan
G1	Public parks									
G11		Comprehensive parks								
G12		Community parks								
G13		Specialized parks								
G14		Mini-parks								
G2	Green buffers									
G3	Squares									
XG	Attached green spaces									
XG-a		Open attached green spaces								
XG-b		Openable attached green spaces								
XG-c		Other attached green spaces								
Total										

Note The data of pocket parks under the category of "formal green space" is reflected by the statistical indicator "G14 Mini-parks". The data of "informal green space" (including pocket parks under the category of "informal green space") is reflected by the statistical indicator "XG-a Open attached green space"; the data of accessory green space with opening up potentials (such as green spaces in the compounds of institutions and schools) is reflected by the statistical indicator "XG-b Openable attached green space"

the common approaches include land purchase, donation, replacement, and no-cost lease, etc. [29]

Taking the United States, where the land acquisition and transfer mechanisms for green spaces are relatively mature, as the example, land purchase is still the most common way to obtain parklands. Most of the earliest urban parks in US, such as the Central Park in New York City and Druid Hill Park in Baltimore [31, 32], were all developed through the land purchase. In most US cities, the government's approaches on purchasing and reserving lands for park development can obtain widespread support from citizens [33]. Meanwhile, US has also fostered the non-profit organizations like "The Trust for Public Land (TPL)" to protect, purchase, and reserve lands for park development [29] (Fig. 6.4). Such organizations usually purchase and reserve the lands for park development in advance in case that the land acquisition conditions for government are not ready. Once the government is enough capable of acquiring these lands, they sell their on-hand lands to the government, completing the transition process of park development lands including land reservation, storage, transaction, and development. It can be observed that the government's financial support and close cooperation with non-profit organizations provide a strong guarantee for the sustained and stable development of pocket parks in the United States.

Although the lands in Chinese cities are publicly owned, the transfer of land use rights still needs considerable efforts on coordination, involving the works such as land expropriation, compensation, and replacement [34]. It also requires strong support from the government's public finance and relevant policies. For example, in recent years, the governments of Beijing, Shanghai, and Nanjing have formulated specific budgets, plans, or programs to support the development of pocket parks (including street-center pockets and street-side pockets, etc.) [35–37]. However, compared to European and US countries, there are still some deficiencies in the land

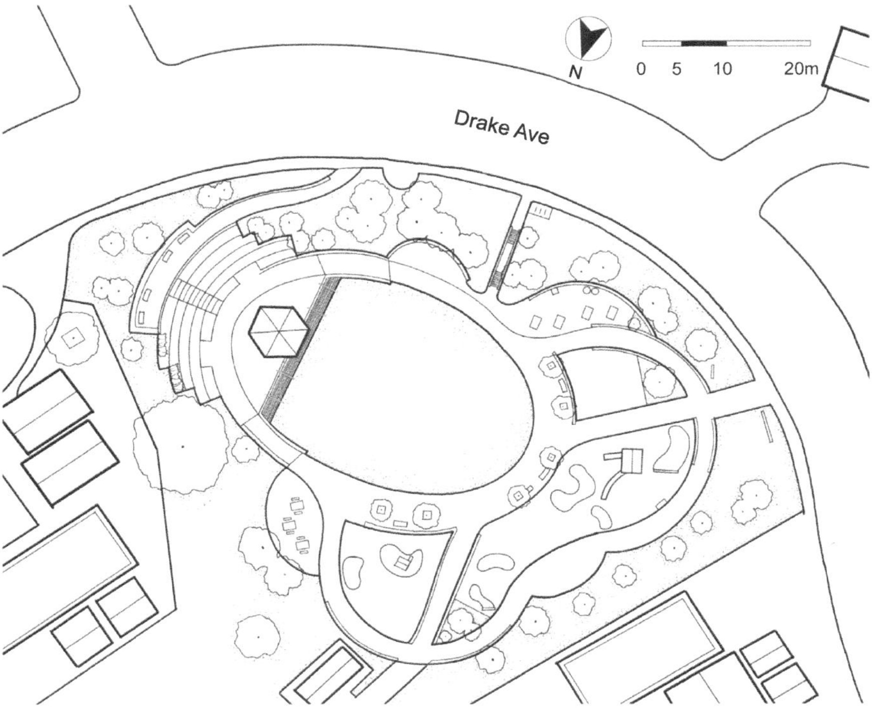

Fig. 6.4 The Trust for Public Land: Plan of George Rocky Graham Park (California, USA)

acquisition and transfer mechanism regarding to urban green space development in China. In particular, the specialized parkland reservation and storage system, like the land bank system in US and European countries [38], that can provide a long-term guarantee for parks' development in the built environment, is still absent.

Meanwhile, the direct support from the government, especially its financial input, also plays an important role in the design, construction, management, and maintenance of pocket parks. The pocket parks that can receive more attention of government, especially the financial investment, usually enjoy high-quality design, construction, and maintenance [39], which is conducive to increasing their recreational attractiveness and service performance.

6.4.2 Indirect Encouragement

Indirect encouragements are mainly employed to support the development of the pocket parks under the title of informal green spaces which do not own dependent land parcels (e.g. attached green spaces). The relevant policies can be mainly aggregated by the policies of development compensation and opening encouragement.

(1) Development compensation

The policies regarding to development compensation mainly encourage developers to reserve adequate ground space for public spaces (including pocket parks) during land development and construction in urban development or renewal, especially by the approaches like floor area incentives and plot ratio compensation (Table 6.4). The development compensation policies first emerged in the *Zoning Resolution- New York City (1961)* of the United States, which proposed to incentivize developers to reserve sites on the ground to provide public spaces such as plazas and green spaces [30]. For example, in the high-density areas of Manhattan, each unit of public space open to street provided by developers at the ground floor can get a reward of ten units of building floor area. This policy paid off well in the subsequent development and renewal of New York, for which there is approximately 4.45 hm2 of public spaces increased in the central Manhattan by 1970 [39] (Fig. 6.5). The success of New York greatly promotes the popularization of such compensation policies to the cities such as San Francisco [40], Chicago [41], Toronto [42], and Singapore [43].

Of Chinese cities, Hong Kong began to learn from New York City's zoning experience as early as the 1970s and explored the adoption of floor area or plot ratio compensation to encourage the development of urban public spaces within the city [44]. In the mainland, Beijing was the first city to explicitly propose the implementation of plot ratio compensation for increasing the open spaces in business and commercial districts in the *Interim Measures for Accelerating the Construction of the Beijing Business and Commercial Center District*, which is promulgated in 2001 [45]. Subsequently, Shanghai, Hangzhou, Nanjing, Wuxi, and other cities in China also adopted similar policies [46]. Development compensation began to be popular

Table 6.4 Types and content of governments' support policies

Categories	Methods	Content	Main applicable objects	Typical forms
Direct support	Budget appropriation	Allocating specific development funds to support the acquisition, storage, planning, design, construction, and maintenance of pocket parks' development lands	Pocket parks under the category of "formal green space"	Mini-parks, etc
	Land acquisition support	Making specific policies to help the land acquisition of pocket parks, especially promoting the land use transition of the inefficient or abandoned lands		
Indirect encouragement	Development compensation	Encouraging development agencies to reserve public spaces in their own leading projects through floor area and plot ratio compensation	Pocket parks under the category of "informal green space"	Streetside green spaces, street-center gardens, etc
	Opening encouragement	Encouraging institutions or private groups to open up previously closed attached green spaces to the public		Time-sharing parks

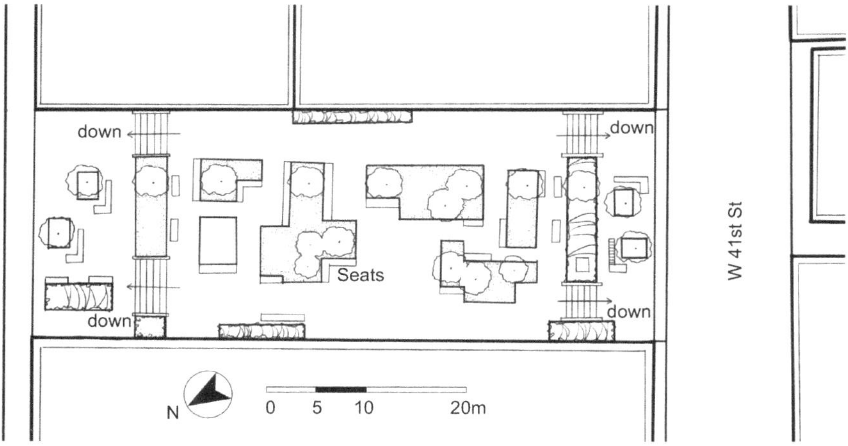

Fig. 6.5 Plan of green space next to 1095 6th. Ave (New York, USA)

in the major cities of China and became one of the main sources of public space growth in high-density areas.

The application of development compensation policies can create numerous opportunities for pocket park installation in the high-density areas. Besides merely development opportunities, there are additional needs of further establishing a classification and control system for the public spaces developed by this compensation mode. With this system, the proper location and spatial scopes for the pocket parks to be installed can be identified, thereby laying the foundation for the refined regulation and control of parks' multiple attributes such as the area, form, facilities, vegetation, spatial perception, and so on [47]. The application of these approaches can significantly enhance the control precision, guaranteeing the service effectiveness and efficiency of the pocket parks.

(2) Opening encouragement

The opening encouragement policy is mainly applied in the scenarios that city's extensive renewal has already been completed. This type of policies focus on opening the existing closed or semi-closed attached green spaces (such as entities' courtyards, compounds, and non-exclusive spaces) to the public through governments' specially-formulated management or coordination measures, which can also achieve the goal of pocket park development [29]. For example, in the US cities like New York, Chicago, and Phoenix, the governmental departments of education and park management jointly cooperated and developed a project titled as "Schoolyard Park". That is to open the internal playgrounds or courtyards of middle and primary schools to the public during after-school hours, using the time-sharing management [48] (Fig. 6.6). In China, Shanghai's *The 15-min Community Life Circle Planning Guidelines (Professional Edition)*, released in 2016, also proposed to encourage to open the attached green spaces of the land parcels of commercial, office, cultural facilities, schools, and residential to the public. It is also recommended that large-sized public spaces attached to cultural and sports facilities and outdoor public spaces such as playgrounds and courts of colleges can be opened to the public. [49] (Fig. 6.7).

The key of the success in the opening and operation of the originally-closed attached green spaces lies in the management and coordination. Of these green spaces, it is relatively easy to open the green spaces attached to the institutional lands with public attributes, such as commercial, office, and cultural institutions, as there are less barriers in the management and maintenance of the post-opened spaces.

Contrarily, the opening of the green spaces within the campuses is much more challenging. However, since the elementary and middle schools are usually proximate to or within the communities, their opening shows more significant effects on responding to residents' daily recreational demands, for which their utilization rate after opening is relatively higher compared to other attached green spaces. According to the development experience of the schoolyard parks in the United States, the involvement of multiple departments often leads to ambiguity in maintenance responsibilities or management standards, making these parks much more difficult to be managed than ordinary neighborhood parks [29]. For example, the education department is typically reluctant to adopt the stricter and more costly

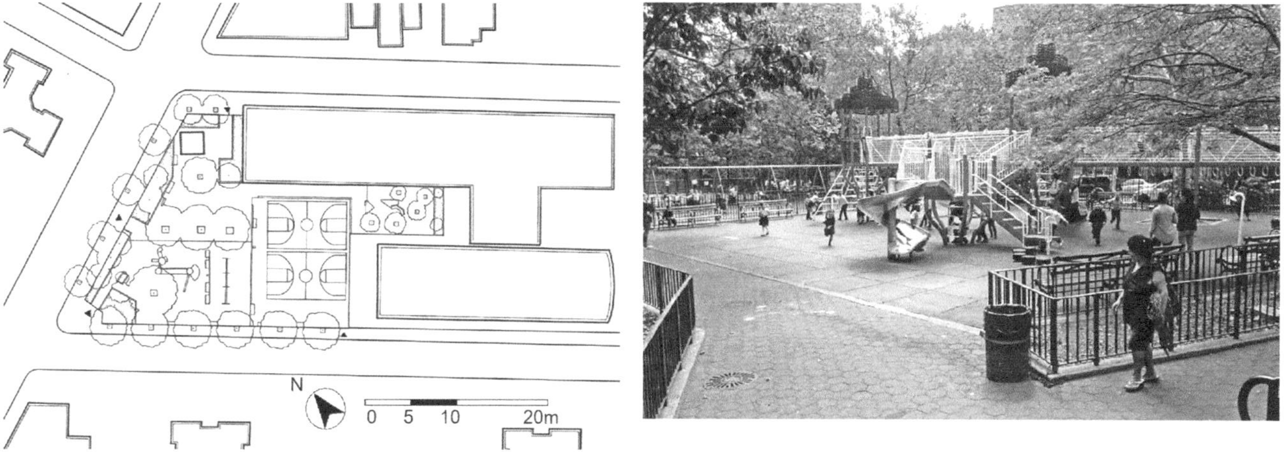

Fig. 6.6 "Schoolyard Park" project: Plan and image of Clark Playground (New York, USA). *Image source* https://nicelocal.com/new-york-city/entertainment/clark_playground/

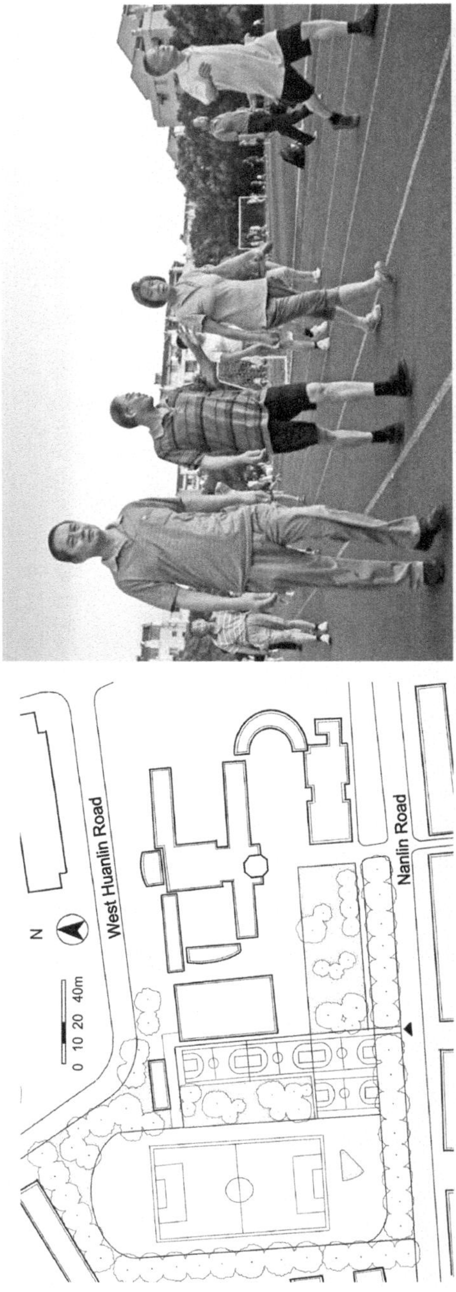

Fig. 6.7 Plan and image of Sanlin Middle School East School open stadium (Shanghai, China) *Image source* https://www.sohu.com/a/167650881_119707

maintenance standards of urban parks in the campuses, while there are also disparities between the recreational demands of the neighboring residents and the playing demands of the students as well [29]. This increases the difficulty of facility configuration and management, thereby requiring specialized and refined coordination to mitigate and handle the possible conflicts in park use.

However, existing practice proved that professional non-profit organizations can effectively contribute to resolve such issues. Taking the Houston of the United States as an example, a non-profit organization named "School Park Program" was specially established in 1983, which effectively promoted the extensive development of school parks [29]. The organization primarily cooperated with the school boards and formulated a sophisticated set of time-sharing control framework and facility configuration standard for the school parks. For example, the campus parks developed by this mode are typically required to configure the combined playing equipment, picnic tables, benches, outdoor classrooms with steps and stages, small courtyards, gravel or granite paths, and native trees. Attributing to the standardized operation processes and refined control, the organization had successfully developed 203 school parks by 2008 [48] (Fig. 6.8).

Due to the unique land ownership system of China, there are abundant enclosed or semi-enclosed attached green spaces that have potentials to be opened. However, due to the lag of the relevant practices, the explorations on the opening forms and management modes of the attached green spaces are just in the preliminary phase in China. The practices on the construction of the incentive system, coordination mechanism, and standardization system of the operation process supporting the opening of attached green spaces are basically absent so far. In this field, the relevant theories and practices in the West can provide useful references for the improvement of corresponding work in China.

6.5 Implementation Guarantee

6.5.1 Design Guidance

The release of the direct support and indirect encouragement policies can create numerous opportunities for the development of pocket parks. However, in this process, to ensure the effective implementation of pocket park planning as well as the service quality and recreational attractiveness of the built-up pocket parks, it still needs the assistance of the refined design guidance and implementation guarantee measures. Taking US cities as an example, the early development compensation policies mainly focused on guaranteeing the quantity of public spaces and successfully achieved this goal after their implementation [29]. However, as the main implemental entities were either enterprises or private institutions such as developers, although many public spaces were successfully reserved and built up, their service content and quality were disassociated from the demands of the users, leading to numerous

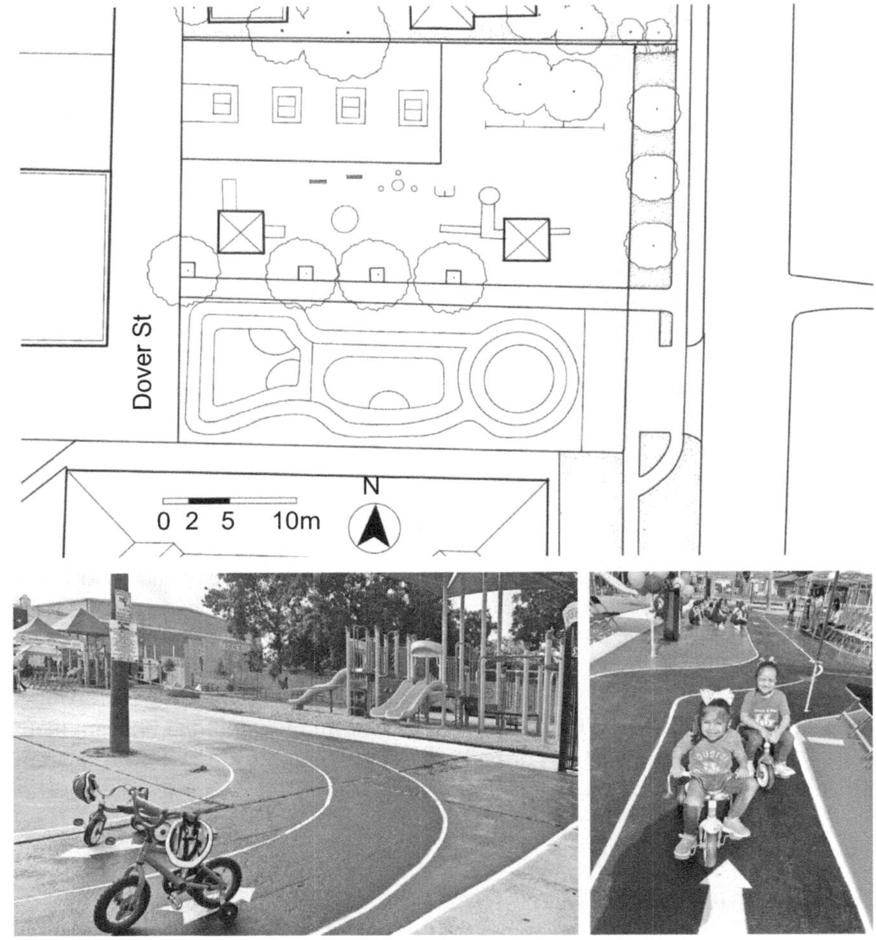

Fig. 6.8 "Schoolyard Park" program: Plan and image of Raul Yzaguirre Schools for Success (New York, USA) *Image source* https://kinder-foundation.org/blog/the-end-of-the-school-year-includes-the-dedication-of-five-new-spark-parks/

inefficiently-utilized public spaces and wastage of land resources [39] (Fig. 6.9). To address this issue, the government explored to regulate and control the various attributes of the compensated open spaces through formulating more detailed guidelines on space design and construction, thereby ensuring the service performance of these spaces after building up [29].

In the *Downtown Plan* of San Francisco formulated in 1985, the types of public spaces were further classified into subtypes, and a complete set of guidelines was proposed for controlling the elements of the public spaces like the materials, seats, vegetation, water features, and sunlight. Only the design of the development projects

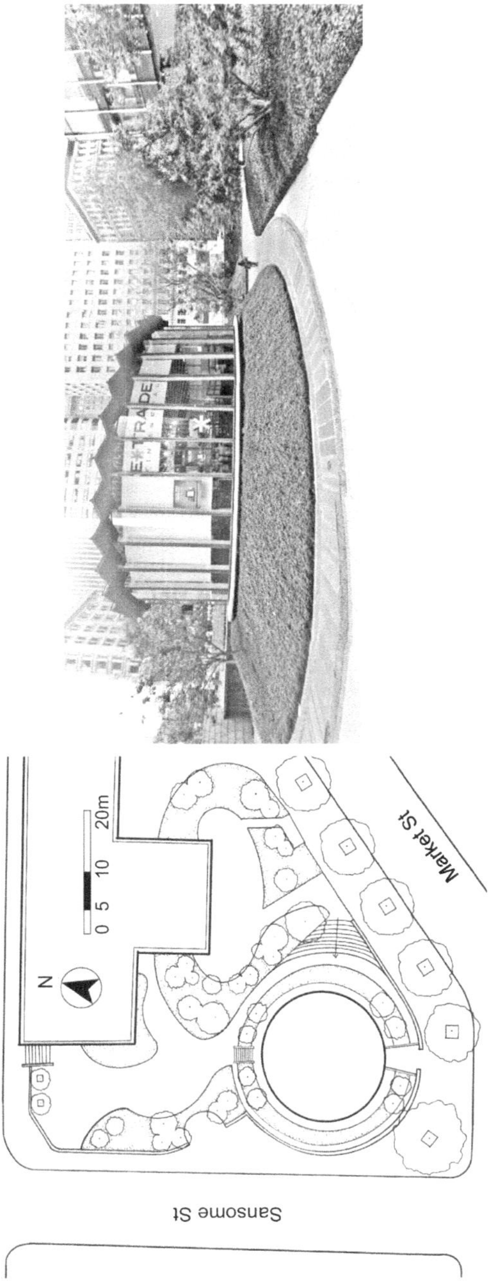

Fig. 6.9 Plan and image of underutilized Zellerbach Square (San Francisco, USA) *Image source* Google street view image

that strictly comply with these guidelines can be approved [50]. Likewise, in the cities such as Cleveland, Chicago, and Los Angeles, it is also mandatory to explain the enhancement effect of the public spaces on promoting the vitality of the surrounding areas before the design is approved [25]. In the *Public Open Space in Private Developments Design and Management Guidelines* released by Hong Kong in 2011, detailed guidance was provided for a total of 14 elements of pocket spaces, that cover the aspects of spatial, perceptual, and landscape planning of such spaces [51].

Building on the existing design guidance modes and frameworks of pocket parks in the world and combining them with the planning thinking of the pocket park groups, a design guidance framework covering both the group and individual levels of pocket parks can be summarized. In this framework, the design guidance at the group level refers to five categories covering 12 control items (Table 6.5) while the design guidance at the individual level encompasses five categories covering 19 control items (Table 6.6).

Table 6.5 Design guidance framework of pocket park groups

Categories	Items	Design guidance
Function	Service structure and division	Group's service definition and type of service structure (horizontal, vertical, or composite); service setting and division of each park
	Main service group	Types and composition of the main targeted users of the park group and individual park
Space	Spatial structure	Spatial structure of the group (monocentric, polycentric, or no center); class of each park in the structure (center or general node)
	Capacity and area	Overall capacity and area of the group; individual capacity and area of each park
	Positioning	Location of individual park (block-corner, block-side, traversing the block, and street-center); orientation of each park
Facility	Recreational facilities	Types, quantity, and capacity of recreational facilities of the park group and individual park
	Other facilities	Types, quantity, and capacity of other facilities of the park group and individual park
Connection	Layout of park links	Location and spatial carriers of the park links
	Forms of park links	Width and section forms of the park links
Style	Overall style	Overall characteristics, identity, and landscaping style of the group
	Correlation	The interrelated spatial forms and elements applied across different parks within a group
	Others	Other design-related factors

Note The specific types of recreational facilities and other facilities can refer to the relevant types in "3.5 Setting of Facilities" of China's national *Code for the Design of Public Park (GB 51192–2016)*

Table 6.6 Design guidance framework of individual pocket parks

Category	Item	Design guidance points
Function	Main service	Main types of service groups; main types of recreational activities suitable for major groups
	Compatible service	Compatible non-dominant group types; compatible types of recreation activities
Space	Area	Land area; the proportion of all types of internal land (green space, hard land, facilities, etc.)
	Shape	Length to width ratio; shape (regular or irregular)
	Boundaries	Boundary types; location and form of street interface; overall enclosure and openness
	Centralized site (main space)	Area and percentage of centralized activity areas; surface materials of centralized activity areas; compatible types of open space activities
	Topography	Site slope; vertical change (site uplift, subsidence or terracing)
Internal constituents	Recreational facilities	Types, quantity, and location of recreational facilities
	Other facilities	Types, quantity, and location of service facilities
	Vegetation	Site retention of plants; additional green plant species, location, forms, colors, etc
Interfaces to surroundings	Entrances	Number, priority and secondary, and location of entrances
	Urban public spaces	Location and connectivity of surrounding sidewalks (or slow-traffic lanes), parking lots, bus stops and subway stations, etc
	External buildings	Coordination with buildings' façades, entrances and exits; setting of green buffers and their width control
	External facilities	Location, types, number, and connectivity of surrounding facilities
Design highlights	Overall style	Individual characteristics, identity, and landscaping style
	Color and material	Color and material of plants, pavements, and facilities
	Light and shadow	Natural lighting, shadow areas and their temporal changes; artificial lighting
	Sightline control	Position and forms of sightline focus; internal and external sightline guidance
	Others	Other design-related factors

Note The specific types of recreational facilities and service facilities can refer to the relevant types in "3.5 Setting of Facilities" of the *Code for the Design of Public Park (GB 51192–2016)*

Besides the detailed design regulations and guidelines formulated and promulgated by government, specific interfaces also need to be set up in the actual planning process to facilitate the adherence and implementation of the detailed planning requirements in the lower-level planning or design. This is particularly crucial for the execution of the service division of pocket park groups for maximizing their overall service performance. To achieve the aforementioned goals, it is necessary to establish a design guidance system at the level of park groups' service (or planning) units. The guidance contents should not only include the factors such as the size, space, boundary, and facilities of each park in the group, but also comprehensively cover the aspects such as the service division, capacity, dominant user groups, and interconnection channels.

6.5.2 Public Participation

Precise response to the detailed demands of the public is a key to the success of the planning, construction, and management of the pocket parks. To ensure the effectiveness and accuracy of this response, it necessitates a thorough public participation mechanism. According to the existing practice of public participation worldwide [52], an effective organization of public participation in the planning, design, and implementations of pocket parks is commonly characterized by four features: fully-participation, adaptability, cultivation, and extensibility.

The feature of fully-participation refers to the establishment of a public participation mechanism that can cover the entire process across the planning, design, and maintenance of the pocket parks. Its specific manifestation lies in establishing the diverse and efficient channels through which the public can express their demands and opinions throughout the entire process and take part in the related work. For example, the United States divides the public participation process during the development of pocket parks into five phases, including preliminary preparation, mission identification, opinion collection, mission implementation, and management and maintenance [3]. The key points of the public participation in each phase cover multiple aspects such as the construction of the public participation framework, design of participation modes, information dissemination and event scheming, construction of the participation assistance organizations, and public education (Table 6.7).

The adaptability is mainly reflected by the customized design of the public participation proposals that can take the unique characteristics of each community into account [3]. For example, the community with a relatively high proportion of the seniors need to establish tailored outreach service teams or organizations, focusing on the on-site information publicity, explanation, and communication assistance, while the communities with a predominance of middle-agers and youths can rely more on the email or social media for information dissemination and communication. In addition to the construction of information interaction channels, the differences in community composition will also impact the selection of personnel for public participation organization, event scheming, and management and maintenance, which also

Table 6.7. Typical processes and highlights of public participation in the development of pocket parks in the U.S.

Phase	Key point	Content
Step one: Getting started	Establish a baseline	● Does the community want a park? ● Is there a potential location that is suitable for a new park and is a park the best use of that site? ● What social, environmental, and health benefits could the park provide?
	Identify partners	● What nearby organizations or agencies support park development? ● What nearby community institutions (place of worship, schools, HOAs) might use the park on a regular basis? ● Are there community centers or business districts that might support the park once built?
	Form an outreach team	● What individuals or groups are important to include on the team? ● What individuals or groups within the community could take on leadership roles within the team? ● What team members are best able to engage with community members about the creation of a new park?
Step two: A shared mission	Community asset mapping	● Identify places near the park where the community gathers, such as schools, libraries, and places of worship. These are good places to make first contact with community members about developing a park
	Pop-up events	● Set up a booth or station organizers around areas in the neighborhood with high foot traffic to peak the community's interest and provide park development and outreach information and gather preliminary feedback
	Community survey	● Surveys can be distributed in person or online to get community feedback about the park project and is a good way to engage a large number of neighbors who may not otherwise be able to participate in meetings or events
Step three: The power of the people	Routine community meetings	● Host regularly scheduled public meetings at a location in the neighborhood that is easily accessible (identified during step two—asset mapping). Meeting activities can include establishing park committees, brainstorming fundraising events, planning for upcoming design charrettes, among others
	Public survey review	● Information gathered from surveys collected during step two can be summarized into a handout or infographic and presented back to community. Public review of this data supports compromise and community collaboration over shared goals for the park
	Design charrettes	● A design charrette is a collaborative public meeting where different options for the park design can be explored and prioritized. These meetings can also be used as forums for the community to select specific park amenities and features

(continued)

Table 6.7. (continued)

Phase	Key point	Content
Step four: The to-do list	Community champions	● The most active community members should be recognized for their work on the project. These leaders are an invaluable resource to the park project. They can be put in charge of organizing park events and are ideal candidates for leaders in step five as they have established ties to the community and a vested interest in the park's success
	Park progress report	● The community should be able to see the progress of a project. This could be shown in a report, checklist, or infographic. Knowing the progress of a project can help the community stay engaged over long periods
Step five: The life of the park	Stewardship program establishment	● If the engagement has been robust, the park committees and community leaders who contributed to the creation of the park maintain their role as park stewards in addition to park users
	Stewardship program diversity	● A park stewardship program should reflect the diversity of the community. Park stewards with different backgrounds will have unique ideas and services to offer. Incorporating a variety of artistic, cultural, and recreational interests allows for park programming that is fresh and most likely to retain the community's continued engagement with the park ● Providing the community with a variety of ways to participate in park stewardship activities opens the door to those who might not be interested in more typical park programming like planting or gardening
	Stewardship program accessibility	● Ensure volunteer and engagement opportunities are known to the community by posting a schedule online and on public bulletin boards within and around the park. This keeps drop-ins as well as veteran park stewards up to date on all opportunities
	Stewardship program empowerment	● Offer ample opportunities with varying levels of physical intensity so any community member, regardless of their physical ability, has a chance to attend. Neither age, experience level, nor physical capabilities should prevent an individual from becoming a park steward or participating in park stewardship activities

Source The Trust for Public Land, 2020 [3].

requires a multi-dimensional and comprehensive design of the public participation proposals.

The cultivation denotes that the willingness of the public participation commonly cannot be fully motivated spontaneously while the participation maturity may not be instantly reached in the initial phase, rather, they need to be continuously nurtured and improved until the goal of full participation can be ultimately achieved [3]. To achieve this goal, it is important to popularize the benefit and value of pocket parks to the public by propaganda while the government and designing agencies can jointly send professional technicians and scheme relevant events or campaigns to educate the public. On the other hand, effective channels need to be built to enable the public to express their opinions and receive feedbacks in the entire process of planning, designing, construction, and maintenance, so as to cultivate the public's sense of ownership. For example, in the *New York City Playground Program* sponsored by the Trust for Public Land (TPL), in order to fully demonstrate the demands of the children, TPL invited the children to develop their own design proposals corresponding to a series of preset questions, which finally generated many creative and implemented design achievements. This mode effectively inspired children's motivation to participate in the park design, during which the children also trained their skills on innovation, communication, and teamwork [48]. In Los Angeles, in order to cultivate and inspire the participation awareness and willingness of the public, the non-profit organization "Los Angeles Neighborhood Land Trust (LANLT)" was established and set the main focus on building the volunteer teams for different communities [3].

The extensibility is mainly relevant to the maintenance and management phase after pocket park construction [3]. To achieve this goal, it requires the competent authorities to transfer some management responsibilities and rights to the local community, such as those related to vegetation planting and maintenance, event scheming and organization, and facility operation and management. This also requires the local community to build a team or organization with sufficient leadership to undertake the responsibility of the long-term management and activity organization within the pocket parks, ensuring that the parks are always in a clean, tidy, safe, comfortable, well-equipped, and effectively running state.

6.5.3 Financing and Management

The funds for pocket park development can be divided into the pre-use and post-use inputs. The pre-use inputs cover all the expenses in the work phases of the land acquisition, planning and design, construction, procedural permits, and other expenses before the completion of the pocket parks [3]. According to the expense types, it can be classified into hard costs (construction-related expenses), soft costs (technical consulting and administrative expenses), and uncertainty costs. The post-use inputs include all the expenses used to maintain the normal running of the pocket parks after they are taken into use, including the utility fees, salaries of the maintenance staffs, replacement cost of the old or damaged constituents, and other costs in park maintenance [3].

There are significant differences in the investment sources between the pocket parks under the categories of formal and informal green spaces. Pocket parks from formal green spaces are usually financed by the greening or park management authority in government. Their development funds source from the government budgets, which are mainly raised by either taxes or bonds. For example, in the United States, the main source of the development funds for park development is the sales tax, property tax, and bonds, while the institutions like non-profit organizations and charities also provide financial support for park development. Even during the 2008

Subprime Mortgage Crisis, the per capita expenditures on green spaces in the cities such as Seattle, Phoenix, and Charlotte still reached to more than $14 [48]. Accordingly, government's financial investment and attentions can build a solid foundation supporting a smooth development of pocket parks.

Pocket parks from informal green spaces are usually not under the jurisdiction of the greening or park management authority. Therefore, their pre-use inputs are mostly undertaken by the landowners' entities, organizations, or institutions. However, there are also some special cases for the pocket parks within certain institutions, such as the school parks. This is because once such pocket parks, which used to be gated and serve for specific groups, are open to the public, it is necessary to enhance their service quality and integrity, thereby incurring additional costs. If the landowners or dominant users undertake the relevant costs, this will bring them with an additional financial burden and reduce their willingness to open the green spaces. Thus, it is necessary to seek extra financial support from external sources or seek opportunities in renovating the management and operational mode of these green spaces [48]. For example, the "School Park Program" in Houston of the United States, a non-profit organization, raises the necessary development funds primarily from charities, corporations, government development funds, and specific committees and control the financial inputs of each park in the range between $75,000 and $100,000 through standardizing the operating procedures and refining the control models. Meanwhile, the schools can cover the daily maintenance expenses of the parks through bake sales, vending machines, miscellaneous sales, and so on [48].

References

1. Shanghai Landscaping & City Appearance Administrative Bureau (2018) Technical guidelines for the construction of street garden in Shanghai (Trial). Retrieved Dec 11, 2024, from https://www.shanghai.gov.cn/nw12344/20200813/0001-12344_55842.html
2. The House of Representatives, Japan (1956) Urban park act. Retrieved from https://elaws.e-gov.go.jp/document?lawid=331AC0000000079
3. The Trust for Public Land (2020) Pocket park toolkit. The Trust for Public Land, San Francisco. Retrieved April 15, 2024, from https://www.tpl.org/resource/pocket-park-toolkit
4. Wu Z, Li D (2010) Principles of urban planning. China Architecture & Building Press.
5. Shenzhen Planning and Land Development Research Center (2016) Shenzhen Green Space System Planning (2014–2030). Retrieved from https://www.sz.gov.cn/szzt2010/wgkzl/glgk/jgxxgk/gtzy/content/post_1413262.html
6. Beijing Municipal Commission of Planning and Natural Resources (2024) Beijing Greenway System Sector Plan (2023–2035). Retrieved from https://yllhj.beijing.gov.cn/zwgk/ghxx/gh/202405/t20240508_3666206.shtml
7. Shenzhen Planning and Natural Resources Bureau (2023) Sector plan for park construction and development in Shenzhen (2021–2035). Retrieved from https://www.sz.gov.cn/attachment/1/1392/1392511/11467133.pdf
8. Nanjing Natural Resources and Planning Bureau (2021) Pre approval public announcement of the regulatory planning and conceptual design scheme for the construction of the tourist center in the Outer Qinhuai River Waterfront Park. Retrieved from https://ghj.nanjing.gov.cn/pqgs/jsxmpqgs/202110/t20211019_3162963.html

9. Nanjing Natural Resources and Planning Bureau (2020) Detailed plan for Nanjing Jiangbei New Area (NJJBh010 Unit). Retrieved from https://ghj.nanjing.gov.cn/ghbz/kzxxgh/202006/t20200601_1900691.html
10. Ministry of Construction of the People's Republic of China (1999) Standard for basic terminology of urban planning. Retrieved from https://www.planning.org.cn/law/uploads/2013/1384002148.pdf
11. Yuan J (2006) Analysis of city green space index. Chinese Landscape Architecture 8:56–60
12. Xinhua News Agency (2019) The Central Economic Work Conference was held in Beijing, and Xi Jinping and Li Keqiang delivered important speeches. Retrieved Dec 15, 2024, from https://www.gov.cn/xinwen/2019-12/12/content_5460670.htm
13. People's Daily (2015) The Central Urban Work Conference was held in Beijing, and Xi Jinping and Li Keqiang delivered important speeches. Retrieved December 10, 2024, from https://www.gov.cn/xinwen/2015-12/22/content_5026592.htm
14. Ministry of Housing and Urban-Rural Development of the People's Republic of China (2019) Standard for planning of urban green space: GB/T51346–2019. China Architecture & Building Press.
15. Ministry of Housing and Urban-Rural Development of the People's Republic of China (2018) Standard for urban residential area planning and design: GB50180–2018. China Architecture & Building Press
16. Ministry of Construction of the People's Republic of China (2017) Guidelines on promoting the work of urban renovation and ecological restoration of the Ministry of Housing and Urban-Rural Development
17. Ministry of Construction of the People's Republic of China (2016) Series standards of national garden city. China Architecture & Building Press
18. Cooper Marcus C, Francis C (1998) People places: design guidelines for urban open space. John Wiley & Sons, New York, NY
19. Yang X, Si M, Hong T (2008) Urban open spaces system planning method research—a case study of Shenzhen City. Planners 24(6):24–27
20. Jiang H, Wu J, Li Z et al (2016) Nanhai open space planning and design guideline toward quality improvement under the inventory growth. Chinese Landscape Architecture 32(12):92–96
21. Zhou C (2015) The public park intensive layout optimization and regulation methods oriented by delicacy management theory. Modern Urban Research 10:47–54
22. Tate, A. (2004). Great city parks. Spon Press, London. Retrieved from https://search.library.wisc.edu/catalog/9910478597202121
23. Whyte, W H, Jr (1980) The social life of small urban spaces. Conservation Foundation, Washington, D.C. Retrieved from https://search.library.wisc.edu/catalog/999547142202121
24. Yang L, Wu Y, Zhou Y (2013) Study on attached green space planning and management in office and residential areas in old city districts—a case study of Changji in Xinjiang. Chinese Landscape Architecture 29(3):55–59
25. SPUR (2009) Secrets of San Francisco: Where to find our city's POPOS—Privately owned public open spaces. Francisco Planning + Urban Research Association (SPUR), San Francisco.
26. Mahmoudi Farahani L, Maller C (2019) Investigating the benefits of 'leftover' places: residents' use and perceptions of an informal greenspace in Melbourne. Urban Forestry & Urban Greening 41:292–302. https://doi.org/10.1016/j.ufug.2019.04.017
27. Rigolon A, Németh J (2018) Privately owned parks in New Urbanist communities: a study of environmental privilege, equity, and inclusion. J Urban Aff 40(4):543–559. https://doi.org/10.1080/07352166.2017.1360739
28. Department for Communities and Local Government (2015) Pocket parks: support for small areas of inviting public space where people can enjoy relief from the hustle and bustle of city streets. London. Retrieved from https://assets.publishing.service.gov.uk/media/5a7f51aaed915d74e6229bd0/Pocket_Parks_Prospectus_Archived.pdf
29. Harnik P (2010) Urban green: innovative parks for resurgent cities. Island Press, Washington, DC. Retrieved from https://search.library.wisc.edu/catalog/9910092652002121

30. Li S, Chen L (2018) 1961 New York City zoning resolution. South China University of Technology Press, Guangzhou
31. Xu H (2003) Urban green system planning overseas. China Architecture & Building Press
32. Xu H (2008) Development and characteristics of American city park system. Huazhong Architecture 26(11):167–171
33. Wagoner C van (2019) The innovative, strategic process of creating Chicago's Millennium Park. Retrieved Dec 15, 2024, from https://cityparksalliance.org/resource/millennium-park-chicago/
34. Shenzhen Municipal People's Government (2015) Several provisions on compensation for land acquisition (conversion), resettlement, and land exchange. Retrieved from https://www.sz.gov.cn/szsrmzfxxgk/zc/gfxwj/szfgfxwj/content/post_11302663.html
35. He G (2018, October 23) Green space per capita increased by nearly 10% in four years—the development of "pocket parks" and "pocket green space" in the abandoned space of street corners and wastelands. *Nanjing Daily*.
36. People's Daily Online (2021, August 30) Pocket park: window view. Retrieved from http://m2.people.cn/r/MV8xXzMxNjk1OTI3XzQxOTg0Ml8xNTg4Mzc5NDQ5
37. Shanghai Observer (2021, August 3) The former road "stone" turned into a pocket park, how many will Shanghai build this year. Retrieved from https://export.shobserver.com/baijiahao/html/385664.html
38. Kildee D, Hovey A, & Vice-President S (n.d.) What is a land bank? *Journal*
39. Cooper Marcus C, Francis C (1990) People places: Design guidelines for urban open space. Van Nostrand Reinhold, New York, N.Y. Retrieved from https://search.library.wisc.edu/catalog/999624093802121
40. American Legal Publishing's Code Library (2024) SEC. 124. Basic floor area ratio. Retrieved Dec 11, 2024, from https://codelibrary.amlegal.com/codes/san_francisco/latest/sf_planning/0-0-0-18102
41. American Legal Publishing's Code Library (2024) 17-4-1000 Floor area bonuses. Retrieved Dec 11, 2024, from https://codelibrary.amlegal.com/codes/chicago/latest/chicagozoning_il/0-0-0-49491
42. City of Toronto (2017) Eco-roof incentive program. Retrieved Dec 11, 2024, from https://www.toronto.ca/services-payments/water-environment/environmental-grants-incentives/green-your-roof/
43. Built Environment Transformation Gross Floor Area Incentive Scheme (2024) Retrieved Dec 11, 2024, from https://www1.bca.gov.sg/buildsg/sustainability/green-mark-incentive-schemes/built-environment-transformation-gross-floor-area-incentive-scheme
44. Qian QK, Fan K, Chan EHW (2016) Regulatory incentives for green buildings: Gross floor area concessions. Building Research & Information 44(5–6):675–693. https://doi.org/10.1080/09613218.2016.1181874
45. The People's Government of Beijing Municipality (2001) Interim measures for accelerating the construction of Beijing Central Business District. The People's Government of Beijing Municipality, Beijing.
46. Hong X, Liu Y, Liang L (2013) Comparative study on the floor area ratio incentive system between China and the United States. In: Collaborative Planning in the Urban Era—2013 China Urban Planning Annual Conference (p. 10). Qingdao, China. Retrieved from https://kns.cnki.net/KCMS/detail/detail.aspx?dbcode=CPFD&dbname=CPFD0914&filename=ZHCG201311002105
47. Ministry of Housing and Urban-Rural Development, PRC (2024) Guidelines for pocket park construction: [2024]214.
48. Harnik P (2010) Urban green: innovative parks for resurgent cities. Island Press, Washington, DC
49. Shanghai Urban Planning and Land Resources Administration Bureau (2016) Shanghai planning guidance of 15-minute community-life circle
50. Wilson Quarterly (1986) San Francisco's Downtown Plan. The Wilson Quarterly (1976–) 10(1): 31–33

51. Hong Kong Port Development Council (2011) Public open space in private developments design and management guidelines. Hong Kong
52. WRI México & American Planning Association (2016) The toolkit for community participation in pocket parks. Retrieved from https://es.wri.org/sites/default/files/pocket_parks.pdf

Chapter 7
Pilot Case I: Mini-Park System Planning in the Central City of Yancheng, China

Unlike the western cities installing pocket park in individual mode, there is much stronger demands for Chinese cities to develop pocket parks in group or system mode due to the higher density and severer scarcity of outdoor recreational opportunities in the built-environment, especially in the old city, historic districts, and CBDs [1]. Accordingly, after 2015, the Ministry of Housing and Urban-Rural Development (MOHURD) of China released a series of standards, Announcements, and guidelines to promote the pocket park (or mini-park) planning and design over the country. In this process, the earlier theories and practices from the western countries provide valuable reference for the relevant practices of China. However, due to the differences in the urban context and development mode of pocket parks, additional innovations on planning theories and methods are required to resolve the unique challenges and fulfill the specific requirements here, especially on the aspects of park group structuring and coordination with the existing green spaces and other elements in the built environment.

In this chapter, we introduced a project titled *Yancheng Urban Park Sector Planning (2019–2030)*, in which the mini-park system planning plays a key role. This planning was jointly formulated by Jiangsu Institute of Urban Planning and Design and our team (Southeast University) in 2019–2022. Considering the limitations of the sector planning on the applicable range of space and land use, this planning only focused on the guidance and control of mini-parks—the pocket parks from formal green spaces. Although the planning of partial pocket parks (from informal green spaces) is not involved, this project also made several new explorations on systematically organizing the service and spatial structure of pocket parks, which are consistent with the contents of the previous chapters.

C. Zhou, *Principles and Methods of Pocket Park Planning and Design*, Urban Sustainability, https://doi.org/10.1007/978-981-96-5801-5_7

7.1 Background

Yancheng is one of the 27 cities in the center of the Yangtze River Delta, located in the eastern coast of Jiangsu Province. According to the seventh national census (November 2020), the population of Yancheng City is about 6.71 million [2]. The central city of Yancheng covered an area of 433 km^2, of which 118.91 km^2 are built-up area, holding a population of 1,099,000 (2016) [3]. The density of the built-up area of the central city is about 10,000 people/km^2 and the density of the core area is more than 30,000 people/km^2 [4], showing high representativeness in the large- and medium-sized cities in China.

Yancheng has made significant progress in green space development and construction in the past 20 years. The city's per capita parkland area rose from less than 3.00 m^2 in the early 2010s to 9.71 m^2 in 2018, accounting for a growth of more than two times [4]. In recent years, oriented by the development concepts and guidelines such as "City Betterment and Ecological Restoration", "Fine Governance", and "Park City", Yancheng has not only paid attention to the growth of green space quantity but also begun to re-examine the rationality of the layout of urban green spaces (especially parks) and their actual service performance. The examination of the park type composition and spatial distribution reveals the serious park service problems in the central city, such as imbalanced composition of park types, uneven distribution in space, and poor accessibility. Of all types of parks, the area of the large-sized comprehensive parks accounts for more than half, most of which are located in the periphery of the built-up area, leading to weak accessibility for most residents. The total area of community parks and mini-parks that served for residents' daily recreation, only account for 7% of parks' total area [4], proving a severe imbalance between the supply and demand of outdoor recreational resources in most densely populated areas (Fig. 7.1).

Correspondingly, Yancheng set the development goals of "citywide parkilization", "centering on human needs and improving livability", and "building park streets and park communities". In 2019, it sponsored the planning project titled as the *Yancheng Urban Park Sector Planning (2019–2030)*, which aims to further optimize the parks' spatial layout, service performance, and allocational equity and rationality through the implementation of this planning. Due to the high density and scarcity of green spaces in the central city, optimizing the spatial distribution and services of parks in the central city was set as the focus of the planning.

However, given the high population density and development intensity of the central city, it is of low feasibility to supplement large urban parks. Therefore, the main object of this planning was centered on the mini-parks with tiny size, aiming to enhance the effectiveness of the park service through the mini-park installation and park group construction.

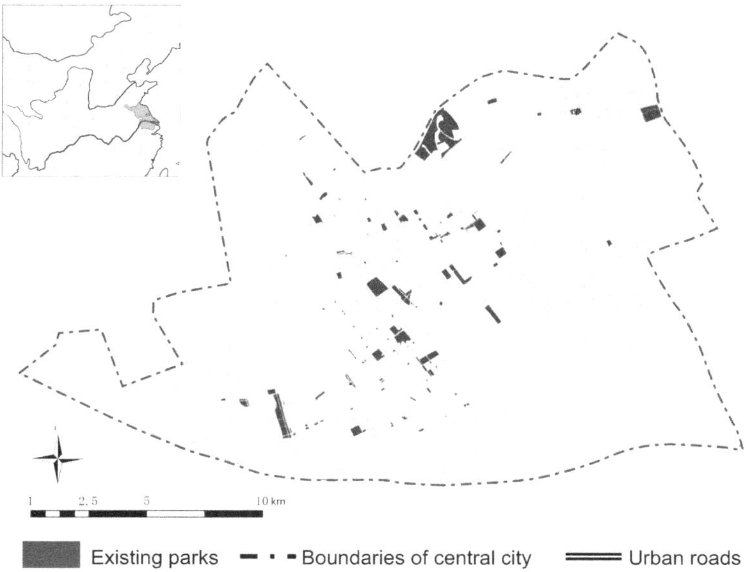

Fig. 7.1 Distribution of public parks in the central city (Yancheng, China)

7.2 Key Strategies

7.2.1 Adjustment of Planning Modes and Indicators

In the preliminary phase of this planning, efforts were made to convert the "system construction" thinking that is developed during rapid urbanization into the "system optimization" thinking that is more adaptive to the current background of urban renewal. This conversion also requires for more precise and efficient methods to comprehensively analyze the diverse factors like the spatial distribution of residents' demand, differences of various groups, and current development conditions. Meanwhile, more planning indicators focusing on the measurements of the quantity, accessibility, and equity of park service were introduced and applied in this planning. The goal is to employ mini-parks as the catalyzer to promote the overall quality of the park service system as well as the urban renewal process.

7.2.2 Construction of Flattened Recreational Service System

Due to the high population density and limited land resources in the planning area, this planning attempted to break up the traditional park hierarchical system by proposing a flattened recreational service system composed by "conventional parks (comprehensive parks, residential parks, and some specialized parks) + mini-parks", aiming to maximumly integrate the existing resources by enhancing the collaboration and complementarity of different types of parks to finally alleviate the serious contradiction between the constrained supply and concentrated demands of daily recreational spaces. According to the features of daily travel of residents and the accessibility standard of green space focusing on 5-min walking distance outlined in the *Guidelines on promoting the work of Urban Renovation and Ecological Restoration of the Ministry of Housing and Urban-Rural Development* [5], 300 m was set as the service radius for green spaces in the daily recreation service system of this planning.

7.2.3 Refinement of Planning Units

To achieve a balanced distribution of green spaces as well as a precise control and regulation, the entire planning area was divided into 178 units, with an average area of about 66 hectares, according to the administrative boundaries of the existing communities and the trunk roads (Fig. 7.2). Of these units, 154 units own residential land, which are taken as the planning units for mini-park installation. Concurrently, with the assistance of the community neighborhood committees, the population and age group composition [including children (0–10 years old), youths (10–30 years old), middle-agers (30–60 years old), and elderly (60 years old and above)] of each planning unit were obtained.

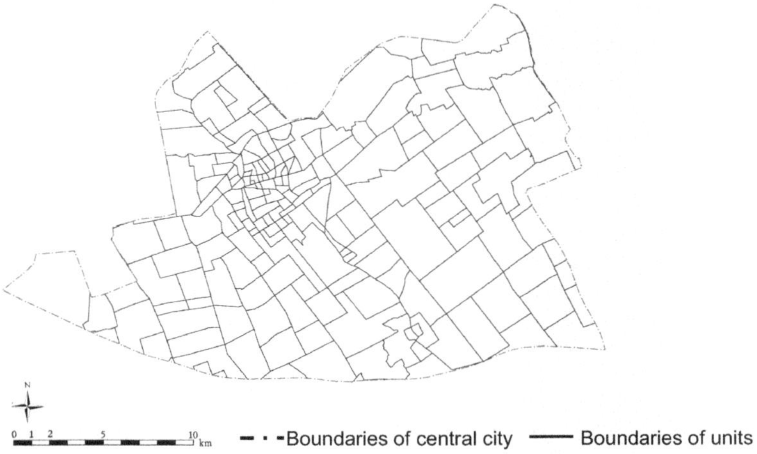

Fig. 7.2 Planning units of the central city (Yancheng, China)

7.2.4 *Establishment of a Planning Framework Based on Supply-Demand Fit*

Given the diverse influencing factors of pocket parks' site selection and service provision [6–10], a planning framework oriented by supply-demand fit was established based on the supply-demand theory of urban public resources [11, 12], facilitating an effective and integrate analysis and control of various factors affecting parks' site selection and service provision. Within this framework, influencing factors were categorized into the ones of the demand-side and supply-side, which are quantitatively and qualitatively analyzed to provide supports for mini-park site selection and service provision.

7.3 Roadmap

According to the planning requirements of Yancheng and the key points of the technical framework listed in Sect. 5.2, the technical framework of this planning contains five main tasks: setting objectives, identifying supplies and demands, comparing supplies and demands, proposing, and validating.

7.3.1 *Setting Objectives*

Since mini-parks do not serve separately in green space system, the examination of their service effectiveness must be integrated into the overall effectiveness of the entire park service system from a comprehensive perspective. Consequently, the objective setting for mini-park planning should also base on their contribution to the service indicators of all the parks. Therefore, besides the planning indicators regarding to park quantity, accessibility, and equity, we also used the indicator of contribution index of mini-parks to examine the contribution of mini-park installation to the entire park system.

After a comprehensive consideration of the existing factors affecting the site selection and service distribution of mini-parks, we set the qualitative and quantitative objectives from the aspects of quantity, accessibility, and equity. The qualitative objectives aim to clarify the parks' development level, while the quantitative objectives aim to use the quantitative measures to achieve the qualitative objectives. Considering the constraints and implemental difficulties of the traditional "one-size-fits-all" quantity-based control mode [13], we explored a more flexible and adaptable control mode featured as "quantity + effectiveness", establishing a multi-dimensional control indicator system to support the formulation and implementation of the planning objectives.

The existing studies and development practice of pocket parks demonstrates that their attributes of the decentralized distribution and the proximate provision of park services can significantly improve the recreational service quality in high-density areas and enhance the green space accessibility [14, 15]. According to these characteristics, the primary contribution (contributing to at least 50% of the increment of the relevant indicators) of mini-parks in this planning was put on the enhancement of recreational opportunity, accessibility, and allocational equity. Meanwhile, the increment of the green space area indicators is listed as the auxiliary contribution (contributing to no less than 25% of the increment of the relevant indicators) (Table 7.1).

7.3.2 Identifying Supplies and Demands

7.3.2.1 Demand Identification

154 planning units were obtained through the unitization and division of the planning area. To further analyze the demand distribution within each planning unit, we first calculated the residential floor area of each residential land parcel within the unit by integrating current land use data with the outlines and the number of floors of the residential buildings, collected from Baidu Maps. Second, based on the proportion of the residential floor area of each parcel in the total residential floor area of the planning unit, the population of the planning unit was distributed to each residential land parcel. This approach refined the spatial granularity of the demand analysis from the community (or regulation unit) level into the land parcel-level (Fig. 7.3).

Meanwhile, we divided the residents into different subdivision groups based on the social attributes. Limited by the accessible data and considering the results of relevant studies proving that there are significant differences in the types and intensity of the daily recreational demands among different age groups. The team used the age attribute to divide the resident groups. According to the age group composition of the demographic statistics in community level, the residents were divided into four age groups including children (0–10 years old), youths (10–30 years old), middle-agers (30–60 years old), and seniors (more than 60 years old) (Fig. 7.4).

The demand intensity of daily recreation across different age groups was identified through a joint adoption of the online and offline questionnaires. The online questionnaire is the primary means for data collection while the offline questionnaire is specially carried out towards the groups of children and seniors, who showed a lower proportion in the recycle of the online questionnaires.

Given that the scarcity of the existing recreational spaces in the high-density environments would limit the actual visit frequency and duration [16], the expected frequency of park visits and duration of each visit is more closed to the real demand of each group and is considered to be more appropriate to be collected in the questionnaires. This data was used to calculate the total expected weekly-park-visit-time,

Table 7.1 Qualitative and quantitative objectives for mini-park installation

Category	Qualitative objectives			Quantitative objectives			
	Existing conditions	Overall objectives	Mini-parks' contribution	Indicators	Existing conditions	Overall objectives	Mini parks' contribution objectives (%)
Quantity	Insufficient park area and recreational opportunities in high density areas	Significantly increasing recreational opportunities and moderately increasing in park area	Primarily and secondarily contributing to the increment of opportunity and area, respectively	Total area of parklands (hm^2)	792	≥1200	≥25
				Number of parks	112	≥400	≥50
Accessibility	Weak accessibility led by the distributional disparity of recreational demand and supply	Significantly improving the accessibility of green spaces for daily recreational service (5-min walking distance)	Primary contribution	Park daily service coverage ratio of residents (%)	24.3%	≥85%	≥50
				Park daily service coverage ratio of residential lands (%)	28.0%	≥85%	≥50
				Proportion of communities with higher than 90% of park daily service coverage ratio (%)	18.1%	≥80%	≥50

(continued)

Table 7.1 (continued)

Category	Qualitative objectives			Quantitative objectives			
	Existing conditions	Overall objectives	Mini-parks' contribution	Indicators	Existing conditions	Overall objectives	Mini parks' contribution objectives (%)
Equity	Severely unequal distribution of park areas and recreational opportunities	Significantly improving the equity of recreational opportunity distribution and moderately improving the equity of park area distribution	Primarily and secondarily contributing to the improvement of distributional equity of opportunity an area, respectively	Gini coefficient of public park area per capita (based on planning units)	0.90	$\leq$0.8	$\geq$25
				Gini coefficient of park number per capita (based on planning units)	0.80	$\leq$0.6	$\geq$60

Note Parks offering daily recreational service include comprehensive parks, residential parks, some specialized parks, and mini parks

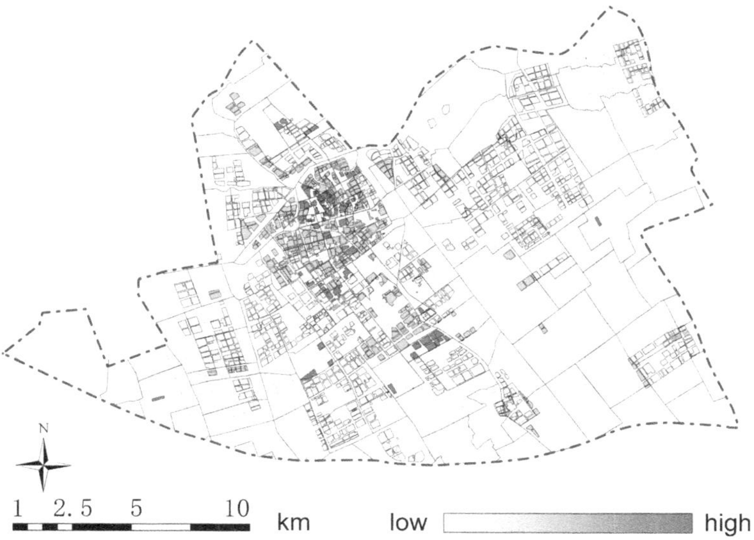

Fig. 7.3 Distribution of demand intensity of residential parcels (Yancheng, China)

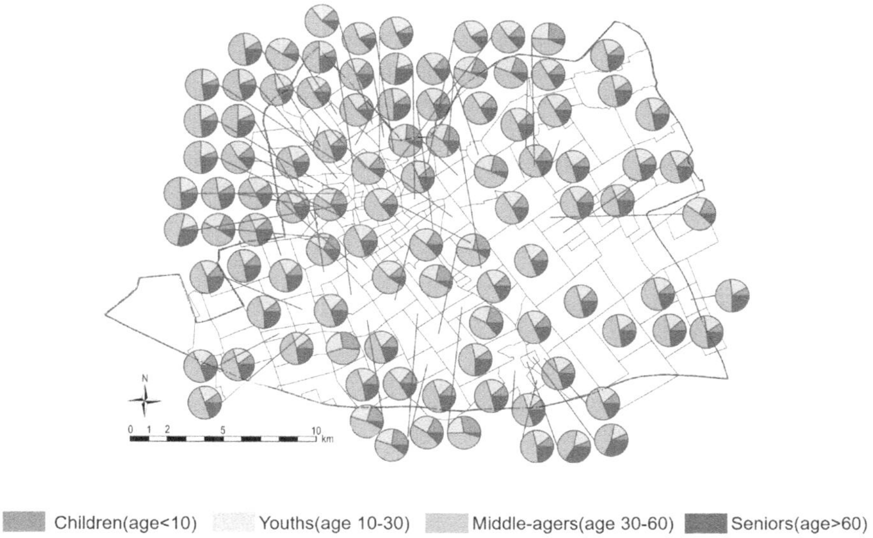

Fig. 7.4 Residents' age structure of each unit (Yancheng, China)

serving as the primary basis for indicating the recreational demand intensity of each age group.

Finally, a total of 532 questionnaires were collected. After excluding those from outside the central city of Yancheng through IP address identification, 343 valid questionnaires remained. These included 55, 123, 93, and 72 questionnaires from children, youths, middle-agers, and seniors, respectively. Through the statistics of the data regarding to the weekly-park-visit-time of various groups, it is found that seniors' expected weekly-park-visit-time spent is the longest, which is 2.5 times longer than that of the youths, whose expected weekly-park-visit-time is the least. The expected weekly-park-visit-time of children and middle-agers ranked second and third, respectively. The expected weekly-park-visit-time of the middle-agers (the largest group of the four age groups) was set as the standard coefficient (valued as 1.0), which is used to derive the recreational demand intensity coefficients of the other three groups by comparing their expected weekly-park-visit-time (Table 7.2).

Combining the population data of each residential land parcel, the composition of age groups, and their demand intensity coefficients, the total demand intensity of recreational service of each residential land parcel was calculated. Then, the buffer zone method was applied to spatially project the demand intensity of each residential land parcel onto its surrounding 300 m areas, by which the spatial distribution condition of the demand intensity was quantified, spatialized, and visualized (Fig. 7.5).

Through this approach, the different groups' demand intensity can be comparable and summable, enabling a citywide and fine-grained analysis on the distribution of the demand intensity (Table 7.3). For example, although both the old town center in the north and the new town center in the south possessed relatively high density, the old town shows significantly higher recreational demand intensity due to the

Table 7.2 Characteristics of recreational demand of the four age groups in the central city of Yancheng

Age groups	Proportion of total population (%)	Proportion of valid questionnaires (%)	Expected frequency of park visits (times/week)	Expected duration per visit (hours)	Total time in park visit per week (hours)	Demand intensity coefficient
Children (aged 0–10)	11.48	16.03	3.2	1.00	3.2	1.2
Youths (aged 10–30)	20.24	35.86	3.5	0.75	2.6	1.0
Middle-agers (aged 30–60)	49.78	27.11	4.0	0.67	2.7	1.0
Seniors (aged >60)	18.48	21.00	4.5	1.50	6.8	2.5

Fig. 7.5 Distribution of demand intensity in different areas (Yancheng, China)

higher proportion of the seniors as higher coefficient of the seniors' demand intensity contribute to a more increment in the demand density quantification.

7.3.2.2 Supply Identification

The quantitative analysis of the factors of the supply-side can be categorized into the measurements of the positive and negative factors in the built environment [8–10, 17, 18] while the qualitative analysis was carried out in the decision-making of the proposing phase. According to the classification of the positive and negative factors affecting pocket park installation and service in Sect. 5.4.2 and the accessible conditions of the data of this project, we quantified the supply condition focusing on the support effect of recreation-related facilities, connection effect of slow-traffic network, and competition effect of existing parks. According to the effect mechanism of the aforementioned factors and the relevant analytical models, the support effect of recreation-related facilities and connecting effect of slow-traffic network were measured by the kernel density model, while the competition effect of existing parks was quantified by the gravity model (Table 7.5).

The measurements of the recreation-related facilities in the central city of Yancheng were carried out focusing on the four types of facilities: restaurants, retail, public toilets, and centralized bicycle parking. The location information of these

Table 7.3 Principles and approaches of the analysis of supply-demand factors

Supply-demand sides	Factors	Principles	Analytical approaches	Data sources
Demand-side	Population density	Higher population density generates higher demand intensity of recreational service	Density statistics	Demographic statistics and land use data
	Group demand intensity	The difference in the weekly-park-visit-times across different groups result in varying levels of recreational demand intensity	Sample survey	Online and offline questionnaires
Supply-side	Support effect of recreation-related facilities	Higher density and proximity of recreation-related facilities produce stronger support effect to parks' recreational service	Kernel density (point density) model	Baidu map POI data
	Connection effect of slow-traffic network	Higher density and proximity of the slow-traffic network produce higher connection effect	Kernel density (linear density) model	Urban greenway network and road system data
	Competition effect of existing parks	Parks with larger area and higher proximity produce stronger attractiveness to residents, intensifying the competition effect	Gravity model	Urban land use data

types of facilities are collected from the POI data of Baidu Maps. Since there is no specific type of centralized bicycle parking in the POI data, the location data of bus stops was collected as a substitute because they proved strong association with the bicycle parking in spatial distribution. The slow-traffic network was extracted mainly based on the data of greenway network and urban road system, while the data of existing parks were primarily extracted from the land use data.

The results showed that the areas with strong support effects of the recreation-related facilities were mainly located in the north and center of the old city which was extended to the north and south along the main road (Fig. 7.6). The areas with strong connection effects of the slow-traffic network were mainly located in the center of the northern district and southern district, where linear open spaces along the roads and rivers concentrated (Fig. 7.7). The areas with stronger competition effects of the existing parks were distributed in the transition areas in the northern and southern districts as well as the fringe areas of the central city (Fig. 7.8). By standardizing and superimposing the three effects, the suitability of mini-parks' development environment (at the supply side) across the central city was derived and spatialized (Fig. 7.9).

Table 7.5 Calculation and mapping methods of supply-demand factors

Supply-demand sides	Measurement indicators	Calculation models	Formula and mapping methods
Demand side	Park demand intensity (PDI)	Scale coefficient superposition model	$\mathrm{PDI}_x = \sum_{i=1}^{n}(D_i\mathrm{Co}_c \times \mathrm{Rc}_i + \mathrm{Co}_y \times \mathrm{Ry}_i + \mathrm{Co}_m \times \mathrm{Rm}_i + \mathrm{Co}_s \times \mathrm{Rs}_i)$ PDI_x is the demand intensity of point x; D_i is the population intensity of parcel I; n is the number of residential lands within 300 m of point x; Co_c, Co_y, Co_m, and Co_s are the demand intensity coefficients of children, youths, middle-agers and seniors, respectively; Rc_i, Ry_i, Rm_i, and Rs_i are the proportion of children, youths, middle-agers, and seniors in parcel i, respectively
Supply-side	Support effect of surrounding recreation-related facilities (IE)	Kernel density model	$\mathrm{IE}_x = \frac{1}{\mathrm{nh}}\sum_{i=1}^{n} k(\frac{x-x_i}{h})$ IE_x is the support effect of surrounding recreation-related facilities for point x, h is the search radius (bandwidth), n is the number of service facilities within the search radius, k(.)is the Kernel function (Gaussian function for this project), (x-x_i) is the distance from facility i to point x
	Connecting effect of surrounding slow-traffic networks (COE)	Kernel density model	$\mathrm{COE}_x = \frac{1}{\mathrm{nh}}\sum_{i=1}^{n} k(\frac{x-x_i}{h})$ COE_x is the connecting effect of surrounding slow-traffic networks for point x, h is the search radius (bandwidth), n is the number of slow-traffic lanes within the search radius, k(.)is the Kernel function (Gaussian function for this project), (x-x_i) is the distance from slow-traffic lane i to point x
	Competition effect of surrounding parks (CE)	Gravity model	$\mathrm{CE}_x = \sum_{i=1}^{n} S_i^{\alpha} \Big/ d_{ix}^{\beta}$ CE_x is the competition effect of surrounding parks at point x, n is the number of surrounding parks, S_i is the area of park i, d_{ix} is the distance between point i and park i, α is the area coefficient, β is the distance attenuation coefficient

Fig. 7.6 Distribution of support effect intensity from recreation-related facilities (Yancheng, China)

7.3.3 Comparing Supplies and Demands

In the supply-demand fit analysis, it is difficult to determine an absolute fit standard due to the subjectivity, diversity, and dynamicity of residents' recreational demands. Therefore, we adopted the relative standard on the measurement of the supply-demand fit. This involved the standardization and grading of the measurement results of the supply and demand conditions. Then, the areas with relatively high grading scores of both supply and demand conditions were prioritized to be selected as the alternative sites for mini-park installation. The measurement results of the recreational demand intensity and the suitability of mini-parks' development environment were divided into five levels: very high, high, medium, low, and very low, with five grading scores from high to low.

On this base, the supply and demand grading scores of each area were superimposed to derive both the sum and difference values of the supply and demand conditions. According to the quantitative principles regarding to supply-demand fit in Sect. 5.3.3, the higher sum value and the lower difference value of the supply and demand conditions indicate the higher supply-demand fit degree, and vice versa, they indicate the lower supply-demand fit degree. Consequently, the results of the supply-demand fit measurement of each area were also divided into five grades from high value to low value (Fig. 7.10, Table 7.6).

Fig. 7.7 Distribution of connecting effect intensity from the slow-traffic network (Yancheng, China)

7.3.4 *Proposing*

7.3.4.1 Land Use Conditions

To control the cost and guarantee the implemental feasibility, we prioritized using the existing open spaces for mini-park installation. Only in case that the available space is extremely limited, converting opening the gated space or demolishing the existing facilities and buildings would be considered as an option. Therefore, our team superimposed the quantitative results of the supply-demand fit with the open space texture of the planning unit and extracted the open spaces with relatively high degree of supply-demand fit as the preferred sites for mini-park installation. Subsequently, the spatial attributes of the sites such as area, form, flatness, slope, orientation, ground attachments, and so on were comprehensively analyzed to derive the construction suitability for intercomparison and decision-making.

7.3.4.2 Development and Implementation Mode

According to the measurements of the construction suitability, we developed three development and implementation modes for mini-park installation in different areas,

Fig. 7.8 Distribution of competition effect intensity from green spaces (Yancheng, China)

Fig. 7.9 Distribution of development suitability from mini-parks (Yancheng, China)

Fig. 7.10 Potential mini-parks layout in the central city (Yancheng, China)

Table 7.6 Results of sum and difference values based on supply-demand fit

Level	Fit degree	Sum value	Difference value
1	Very high	≥8	≤1
2	High	≥7	≤2
3	Medium	≥6	≤3
4	Low	≥5	≤4
5	Very low	≥5	≤4

namely, "converting the hard ground", "inserting into the existing greenery", and "demolishing and reconstructing" (Figs. 7.11 and 7.12). The first two modes are applicable to the scenario of installing parks in the existing open spaces, while the last mode is applicable to the scenario of installing parks through land use conversion. To control the costs and facilitate the planning implementation, the first two modes were set as the primary alternative modes for mini-park installation of this planning, while the third one was used as the secondary alternative mode.

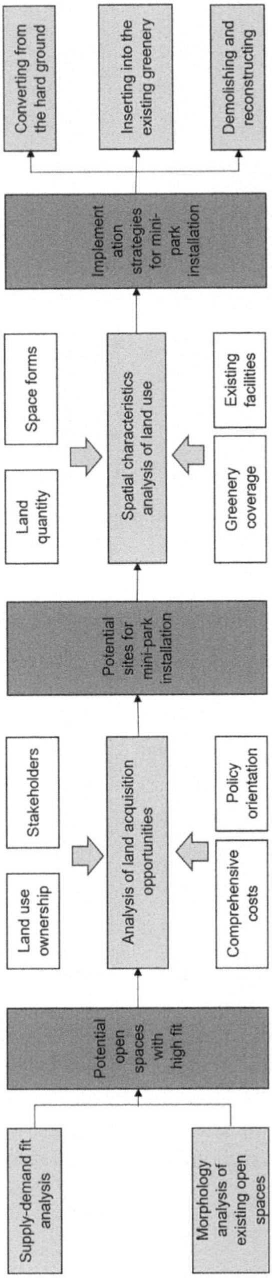

Fig. 7.11 The implementation path of mini-park installation

Fig. 7.12 Existing and newly planned mini-parks in the central city (Yancheng, China)

7.3.4.3 Grouping

Additionally, we also took the factors affecting the collaboration of park group into account in the site selection of mini-parks within each planning unit. Strategically, in order to enhance the adaptivity of the group's spatial distribution to the surroundings, we made great efforts to establish the spatial structure of the park groups, ensuring that it can match with the existing fabric and resident distribution of the planning units. To couple the service of park group with the distribution and composition of the demands of different groups, specific attentions were paid in the differentiated division and flexibility reservation of the park group service. To build an efficient connection network and promote the interaction with the surroundings, the areas with diverse recreation-related facilities around and convenient connection interfaces were prioritized as the park installation sites (Figs. 7.13, 7.14, 7.15 and 7.16).

7.3.4.4 Results

After the above approaches and multiple rounds of coordination and negotiation with the local government and stakeholders, the planning proposal of mini-parks was finally formed. In the proposal, the total area of mini-parks increased from 28.44 hm^2 to 199.01 hm^2, and their total number increased from 54 to 395. Of all the increased mini-parks, 103 parks (30.2%) were proposed to be installed through "converting the hard ground", 193 parks (56.6%) were proposed to be installed through "inserting into the existing greenery", while 45 parks were proposed to be installed through "demolishing and reconstructing" (13.2%) (Fig. 7.17).

7.3.5 Validating

7.3.5.1 City-Level Validation

A comparison between the planning objectives and the relevant parameters of the planning proposal confirmed a successful achievement of the preset goals. Specifically, the increment of the quantity parameters proved that the mini-park installation contributed much more to the increment of park number (79.8%) than that of park area (35.4%). The adoption of the park development strategy focusing on mini-parks in Yancheng's central city brought about nearly 3.8-fold increase in the number of daily recreational parks, greatly exceeding the area increase by approximately 0.6-fold (Fig. 7.18). This also reaffirmed that under the conditions of limited land resource, the development of pocket parks contributes much more significantly to the increment of outdoor recreational opportunities than that of area growth [19, 20]. From the perspective of the effectiveness of green space services, the great increase of recreational opportunities can also alleviate the conflict between the supply and demand of outdoor recreational resources in high-density areas.

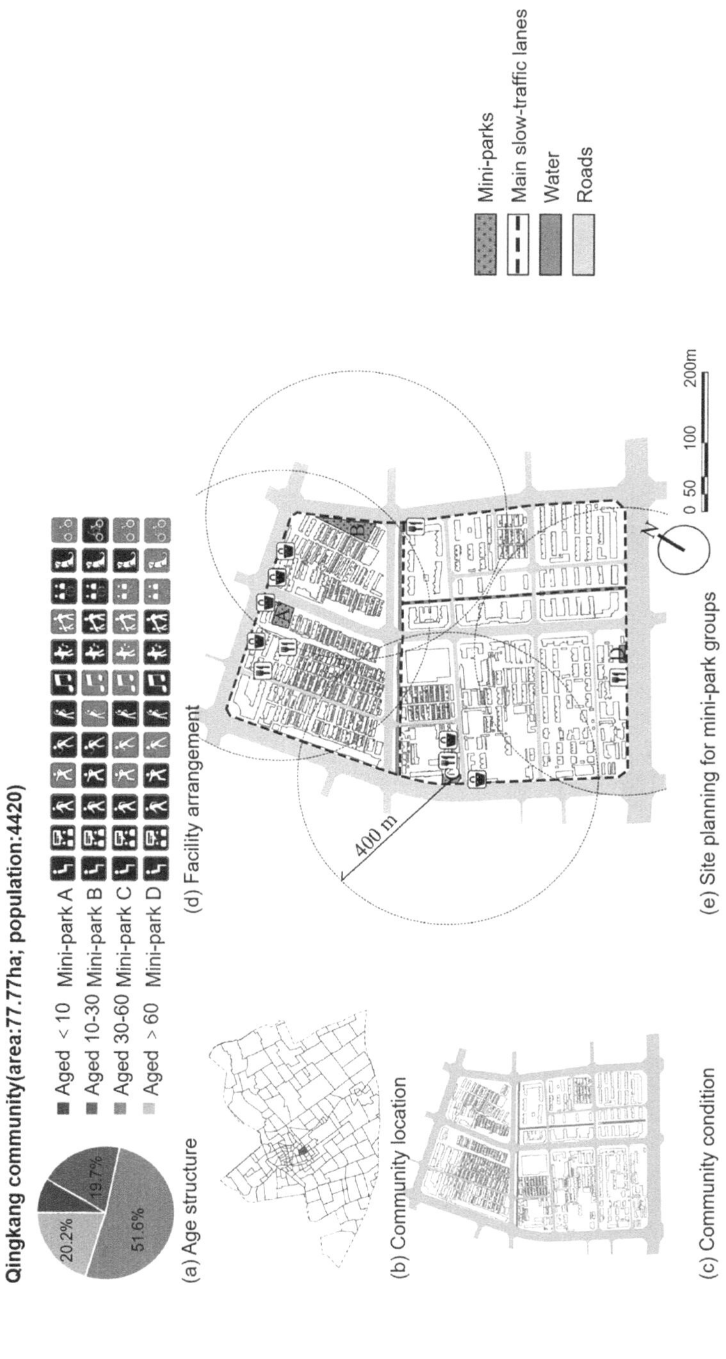

Fig. 7.13 An example of a ring-structured group: Qingkang Community (Yancheng, China)

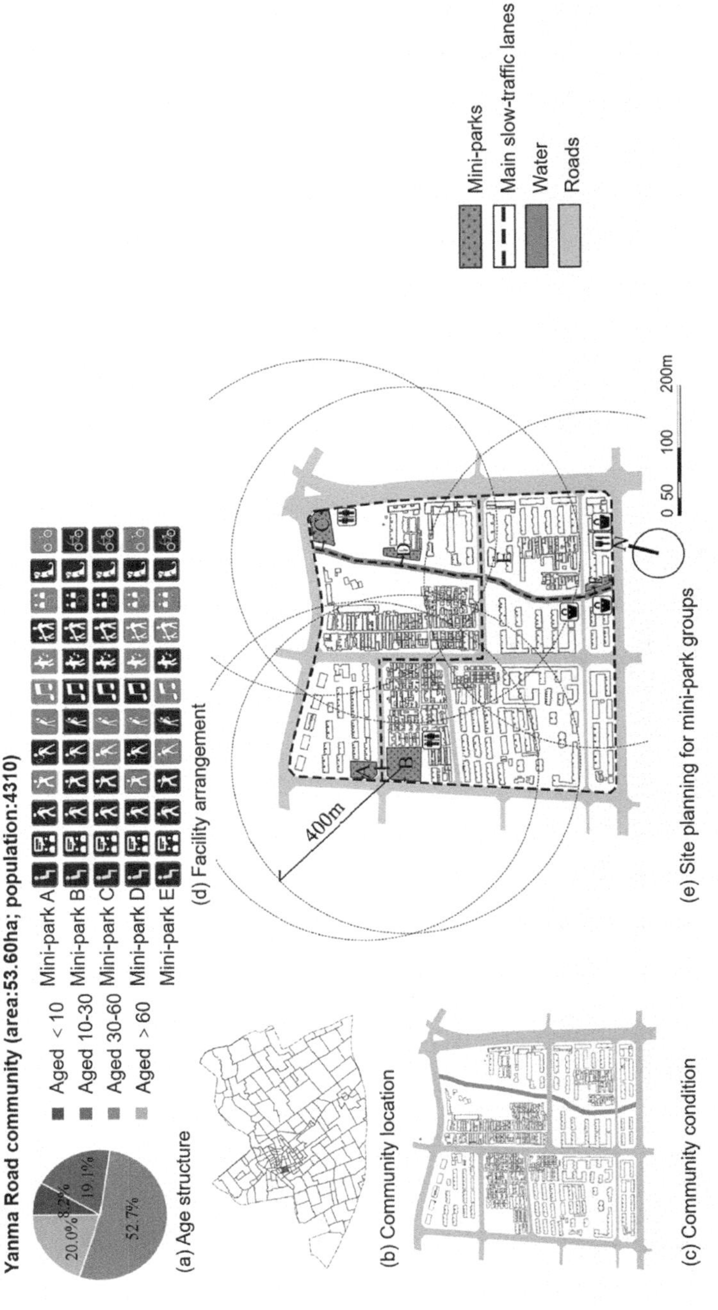

Fig. 7.14 An example of a linear-structured group: Yanma Road Community (Yancheng, China)

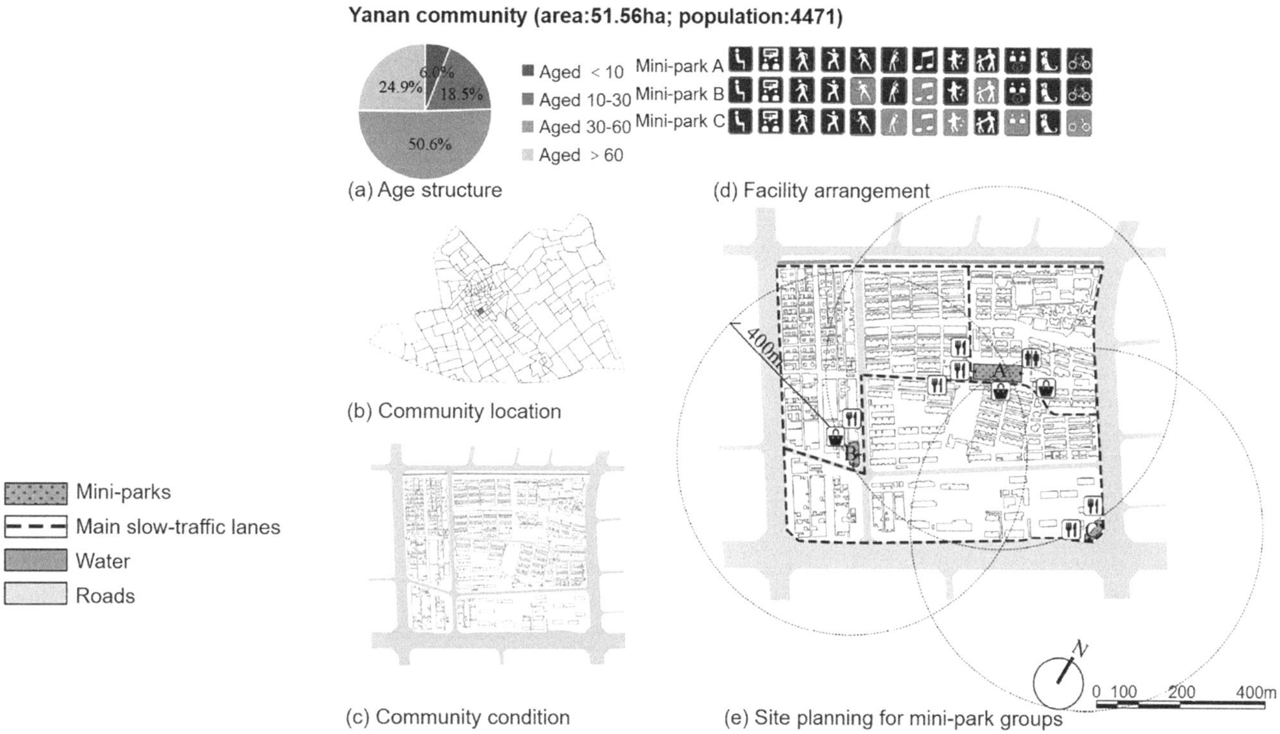

Fig. 7.15 An example of a radial-structured group: Yannan Community (Yancheng, China)

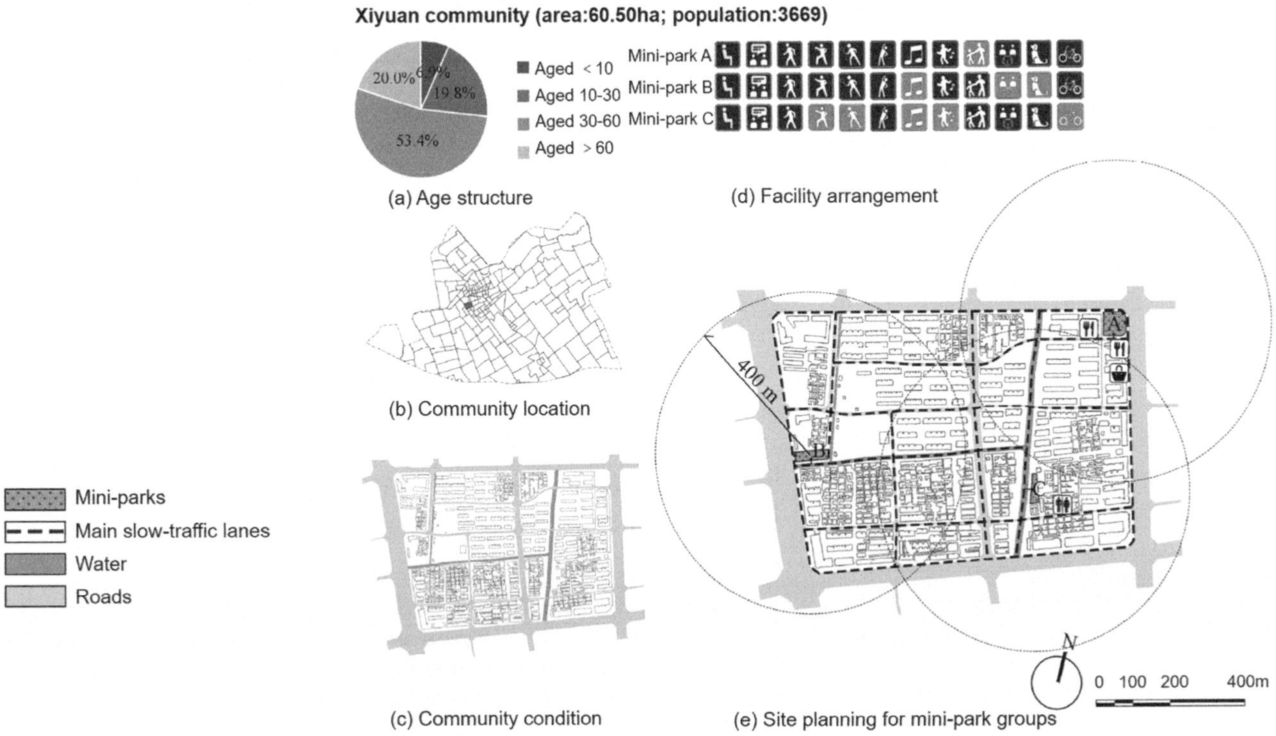

Fig. 7.16 An example of a network-structured group: Xiyuan Community (Yancheng, China)

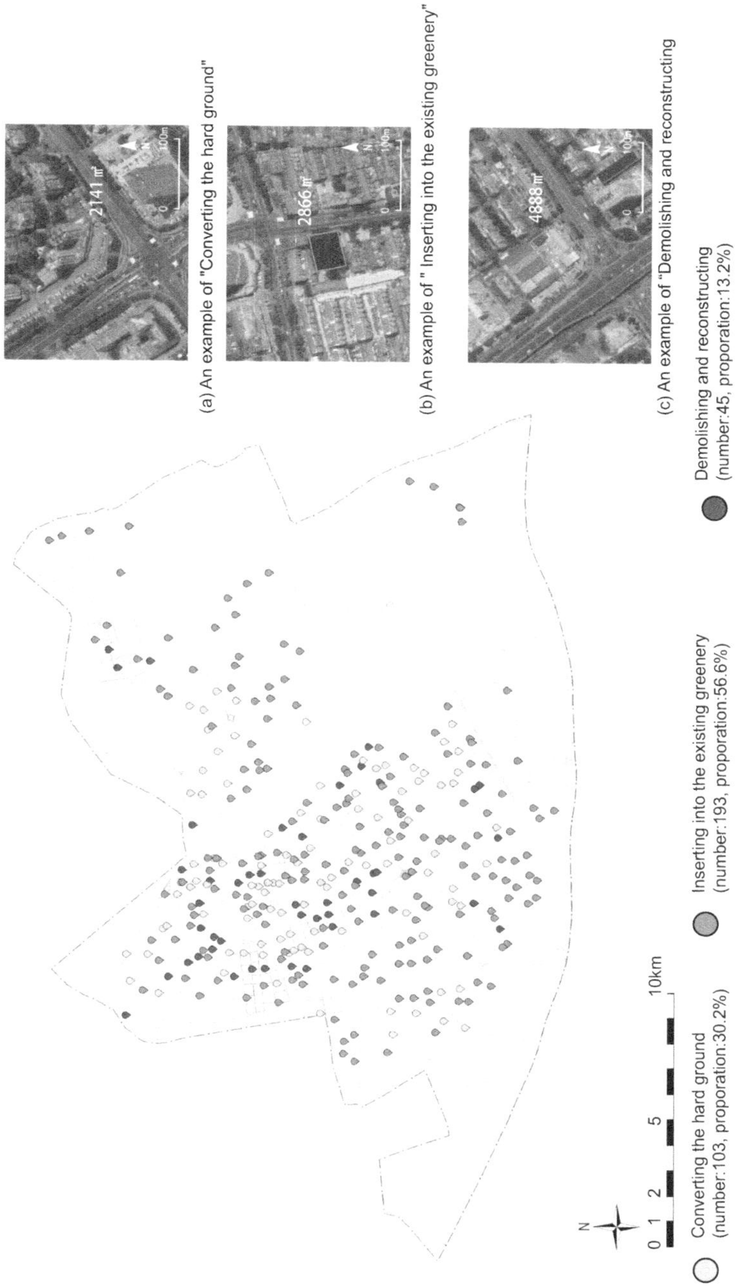

Fig. 7.17 The mini-park installation proposal and typical implemental examples in the central city (Yancheng, China)

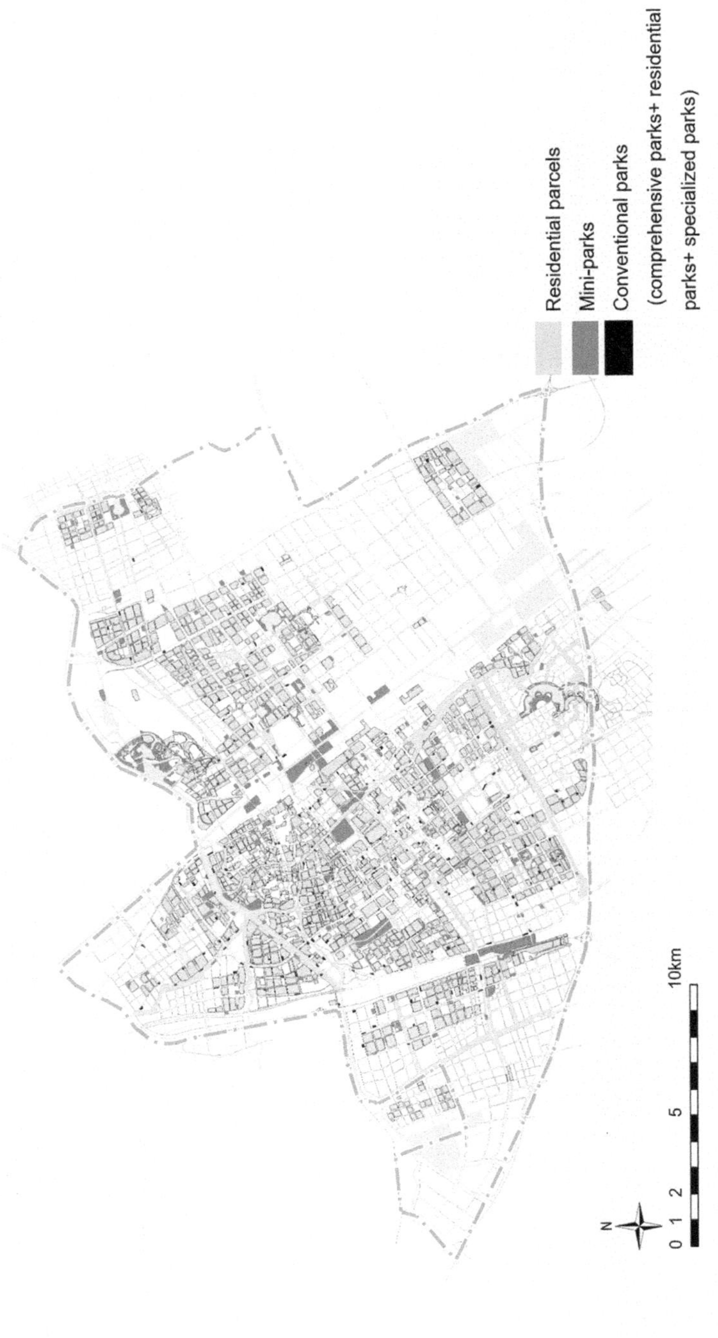

Fig. 7.18 The planning proposal of mini-parks in the central city (Yancheng, China)

Meanwhile, the mini-park installation also contributed to more than 70% of the increment of the accessibility and equity level, acting as the main power to enhance the park service effectiveness. However, there was a significant disparity between the increments of these two parameters. The increment of accessibility reached to more than two times whereas the increment of equity is much lower, for which the Gini coefficient of park area per capita and the Gini coefficient of park number per capita still indicated a severe gap (higher than 0.5) [4] in park distribution among different units after the planning. This proved that in high-density areas, planning measures focusing on mini-parks are insufficient to completely reverse the severe imbalanced distribution of parks across planning units. Nonetheless, comparing the increment of the equity of park area distribution and that of park number distribution, the mini-park installation contributes more significantly to the improvement of the latter one (Table 7.7).

7.3.5.2 Unit-Level Validation

For the park group construction within the planning units, we used the residential land parcels as the analytical object and verified the improvement of the supply-demand correlation, spatial balance, and social equity. The results showed that the supply-demand correlation of park service was improved for most units, in which the number of units that reach to an approximate balance status of supply-demand of park service (park area per capita reach to no less than 2.4 m^2 according to the planning standards of China) increased from 40 to 83. However, due to the high density and land scarcity, the quantity of the parks offering daily recreation service in the city center was still insufficient, but the extent of insufficiency had been mitigated. Accordingly, the planning proposal further recommended an extra construction of greenway networks, rooftop greening, and the opening up of attached green spaces to address this issue.

By analyzing the standard deviation of the park area within the 300 m distance around each residential parcel of the planning units, we further measured the spatial distributional balance of the parks. Results demonstrated that the proportion of the units with a high level of spatial balance accounts for about 70% of the total units. However, there were still some units showing imbalanced spatial distribution of parks, mainly concentrated in the central and eastern boundary areas of the central city. Some of them were spatially constrained in achieving a balanced distribution of parks (Fig. 7.19). Therefore, the additional approaches such as constructing a greenway network, opening gated community green spaces, and opening green spaces attached to surrounding public facilities can be considered to enhance the spatial distributional balance of recreational resources rather than directly installing parks. Furthermore, the distributional imbalance for some units was led by the existence of the large parks. In this scenario, greenway network is prioritized to be constructed to enhance the accessibility and visit convenience of the residents around, aiming to use an efficient connection system between the supply and demand sides to compensate for the spatial imbalance in park distribution.

Table 7.7 Comparisons of key indicators, increase, and mini-park contribution

Items	Indicators	Before planning	Overall objectives	After planning	Increase (%)	Mini parks' contribution objective (%)	Mini-parks' actual contribution after planning (%)
Quantity	Total area of parklands (hm^2)	792	≥1200	1270	60.4	≥25	35.4
	Number of parks	112	≥400	538	380.4	≥50	79.8
Accessibility	Park daily service coverage ratio of residents (%)	24.3%	≥85%	97.7%	302.1	≥50	70.8
	Park daily service coverage ratio of residential lands (%)	28.0%	≥85%	90.1%	221.8	≥50	79.0
	Proportion of communities with higher than 90% of park daily service coverage ratio (%)	18.1%	≥80%	80.7%	345.9	≥50	81.1
Equity	Gini coefficient of park area per capita (based on planning units)	0.90	≤0.8	0.78	13.3	≥25	83.3
	Gini coefficient of park number per capita (based on planning units)	0.80	≤0.6	0.55	31.6	≥60	80.0

Note Parks offering daily recreational service include comprehensive parks, residential parks, some specialized parks, and mini-parks

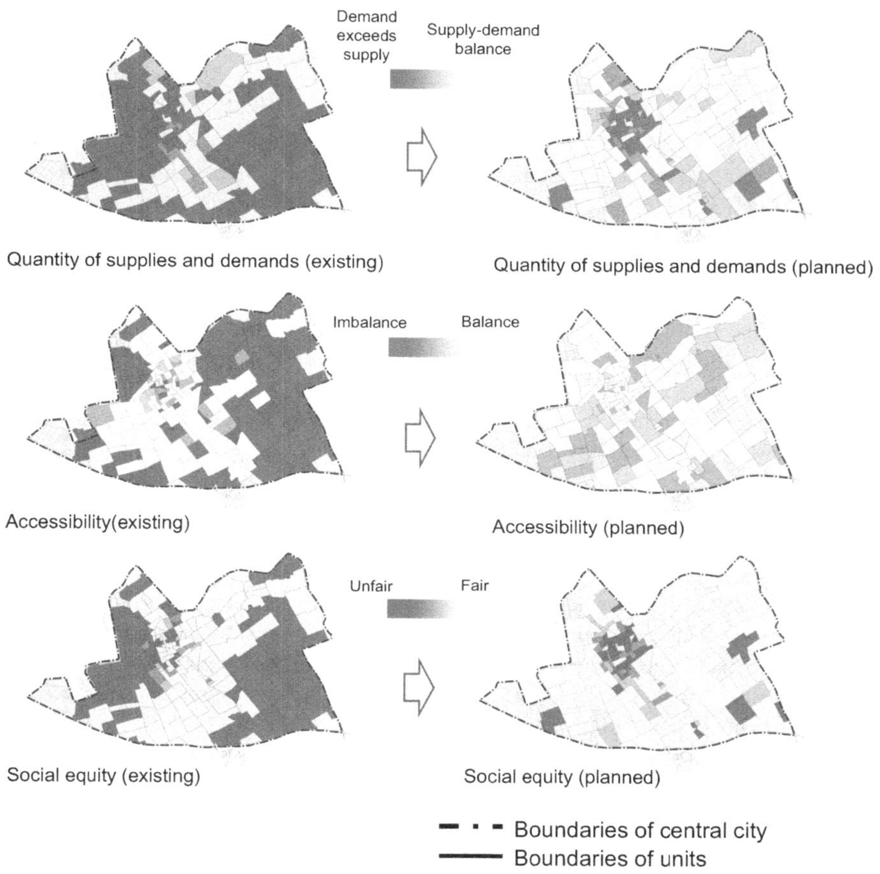

Fig. 7.19 Validation of regulation units (Yancheng, China)

The measurement of the social equity regarding to the park service allocation within the planning units mainly focused on the differences in the degree of the park occupation of various age groups (Gini coefficient). The results showed that the number of planning units with a severe gap in park number allocation (Gini coefficient higher than 0.5) is reduced by 30, indicating the degree of social inequality in the park service allocation is alleviated to a certain extent. However, due to the high saturation of land development in some built-up areas, there are still some communities in the middle and southeast of the city showing severe inequality in park service allocation among different age groups. To mitigate the issues of these units, trails, informal green spaces, and recreational facilities specially serving for the vulnerable groups were recommended to be installed.

7.4 Summary

7.4.1 Characteristics and Effects of the Method

In this planning, we refined the spatial and social granularity in the analysis based on the subdivision of the planning units and resident groups, through which the uneven distribution of residents' demands in different areas and age groups can be integrally deliberated and measured. As reflected in the planning results, park service coverage ratio of "people" is nearly 8% higher than that of "land", while the increment of the former one is almost 80% higher than the latter. This in fact transfers the focus of the park spatial distribution from the traditional "land-land" (parkland-residential land) relationship to the "land-people" (parkland-residents) relationship, aligning with the current planning requirements and orientations under the background of urban renewal and refined governance.

Given the land resource restriction of park development in the high-density environment, the planning laid the focus on supplementing and redistributing the outdoor recreational opportunities. In supply-demand fit degree measurement, the existing transportation and recreation-related facilities in the built environment were integrally considered, together with the competition effect of the recreational service between parks. This constructed a solid foundation for resource integration, cost control, and effectiveness guarantee in the process of mini-park development, through which the mini-park installation can have a chance to achieve the goal of "small efforts with significant impact" in urban renewal.

Due to the challenges in constructing an absolute supply-demand fit standard, the relative fit degree measurement was adopted as a substitutive approach in this planning. This approach can effectively identify the priority sites for mini-park installation with the maximum adherence to the urban realities, for which it shows high applicability in the optimization of the park distribution in current urban renewal process of Chinese cities and similar cities worldwide.

7.4.2 Limitations

There are three limitations in the application of this method. First, given the complexity of land acquisition in the park development of the high-density environment, it is difficult to fully quantify the key factors in land supply. Instead, the strategy used in this planning is to divide the relevant factors into the quantifiable and unquantifiable factors, that is, firstly extracting the priority alternative sites for mini-park installation according to the coupling result of the demand factors and the quantifiable factors of the supply side (e.g. factors regarding to development suitability) and subsequently finalizing the sites for mini-park installation through qualitative analyses of the unquantifiable factors such as the policy, economic, and social factors related to land acquisition. If future studies can establish a comprehensive and objective method to quantify all the factors affecting land acquisition, it will further enhance the efficiency and accuracy of the measurement on supply-demand fit.

Second, as the planning is titled as *Yancheng Urban Park Sector Planning (2019–2030)*, the planning object had to be set on the mini-parks under the category of parklands (formal green spaces). They did not encompass the pocket parks under the category of informal green spaces, which actually play similar roles as mini-parks. Although the method developed in this planning is also applicable to the informal green spaces, this limitation affects the comprehensiveness of the construction of the daily recreational service system and park groups in the planning results.

Finally, in the measurement of the supply-demand fit, only the differences among the age groups were taken into accounts at the demand side. However, existing studies proved that differences also exist among the various groups divided by the other social attributes such as occupations, incomes, genders, and so on. Therefore, further exploration is needed to develop a more multi-dimensional and refined measurement to integrate these differences into the planning analysis and decision-making. Furthermore, due to the limitation of the labor and time, this planning only assessed the demand intensity of different age groups, failing to collect the information on the disparities of demand types of different age groups and integrate them into the supply-demand fit assessment. On the supply side, this planning primarily quantified the influence of the three types of factors (recreation-related facilities, slow-traffic networks, and existing parks). With the proceeding of the relevant studies, other constraints or opportunities for park development that exist in the built environment may be identified and quantified, which can perfect the assessment system of the development suitability in this planning.

References

1. Zhou C, Yang L, Jiang Z, et al (2024) A deep learning approach to identify potential sites for pocket park installation in Nanjing, China. J Urban Technol 1–27. https://doi.org/10.1080/10630732.2024.2402676
2. China Statistics Press (2021) Major figures on 2020 China population census. Retrieved from https://www.stats.gov.cn/sj/pcsj/rkpc/d7c/
3. China Statistics Press (2017) 2017 Yancheng statistical yearbook. Retrieved from http://tjj.yancheng.gov.cn/col/col1779/index.html
4. Zhou C, Zhang Y, Fu L, et al (2021) Assessing mini-park installation priority for regreening planning in densely populated cities. Sustain Cities Soc 67:102716. https://doi.org/10.1016/j.scs.2021.102716
5. Ministry of Construction of the People's Republic of China (2017) Guidelines on promoting the work of urban renovation and ecological restoration of the Ministry of Housing and Urban-Rural Development [S]
6. Boulton C, Dedekorkut-Howes A, Byrne J (2018) Factors shaping urban greenspace provision: a systematic review of the literature. Landsc Urban Plan 178:82–101. https://doi.org/10.1016/j.landurbplan.2018.05.029

7. Hamstead ZA, Fisher D, Ilieva RT et al (2018) Geolocated social media as a rapid indicator of park visitation and equitable park access. Comput Environ Urban Syst 72:38–50. https://doi.org/10.1016/j.compenvurbsys.2018.01.007
8. Harnik P, Green U (2010) Innovative parks for resurgent cities. Island Press, Washington, DC
9. Forsyth A, Musacchio L (2005) Designing small parks: a manual for addressing social and ecological concerns. John Wiley & Sons
10. Rioux L, Werner CM, Mokounkolo R et al (2016) Walking in two French neighborhoods: a study of how park numbers and locations relate to everyday walking. J Environ Psychol 48:169–184. https://doi.org/10.1016/j.jenvp.2016.10.003
11. Wen Z, Zhang S, Yang Y, et al (2023) How does enclosed private residential green space impact accessibility equity in urban regions? A case study in Shenzhen, China. Urban For Urban Green 85:127968
12. Rao Y, Zhong Y, He Q et al (2022) Assessing the equity of accessibility to urban green space: a study of 254 cities in China. Int J Environ Res Public Health 19(8):4855
13. Liu D, Kwan MP, Kan Z (2021) Analysis of urban green space accessibility and distribution inequity in the city of Chicago. Urban For Urban Green 59:127029
14. Harper J (2009) Planning for recreation and parks facilities: Predesign process, principles, and strategies. Venture Publishing, Inc. https://search.library.wisc.edu/catalog/9910467099002121
15. Cooper Marcus C, Francis C (1990) People places: design guidelines for urban open space. Van Nostrand Reinhold. https://search.library.wisc.edu/catalog/999624093802121
16. Jim CY (2004) Green-space preservation and allocation for sustainable greening of compact cities. Cities 21(4):311–320. https://doi.org/10.1016/j.cities.2004.04.004
17. Armato F (2017) Pocket park: product urban design. Des J 20(sup1):S1869–S1878. https://doi.org/10.1080/14606925.2017.1352705
18. Faraci P (1967) Vest pocket parks [R]. Am Soc Plann Off.
19. Kang Y, Fukahori K, Kubota Y (2018) Evaluation of the influence of roadside non-walking spaces on the pedestrian environment of a Japanese urban street. Sustain Cities Soc 43:21–31. https://doi.org/10.1016/j.scs.2018.07.009
20. Nordh H, Østby K (2013) Pocket parks for people—a study of park design and use. Urban For Urban Green 12(1):12–17. https://doi.org/10.1016/j.ufug.2012.11.003

Chapter 8
Pilot CaseII—AI-Based Identification of Pocket Parks' Installation Sites

The analysis of supply and demand conditions can preliminarily identify priority areas for pocket park installation, while it is still challenging to anchor the exact installation location of each pocket park in densely populated areas. This is greatly due to the high density and complexity of the built environment, leading to fragmented spaces and diverse factors affecting the installation of pocket parks. Most of these fragmented spaces are usually part of the existing land parcels or facilities which cannot be demonstrated by the traditional land use data. Meanwhile, potential sites for pocket park installation (PPI) are typically small-sized, dispersedly-distributed, and situated in an environment with diverse constraints, raising higher requirements of the dimensions, breadth, and accuracy for identification task [[1]]. Traditional analytical tools, such as accessibility models, are only capable of measuring two-dimensional spatial information, which is difficult to accurately measure the fine-grained features of site conditions and surroundings. Although the site selection method based on on-site investigation, assessment, and expert experience [[2]] can broaden the measurement dimensions, it requires substantial manpower, materials, and financial resources [[3]], showing great limitations in application scope and judgement comprehensiveness [[4]].

Due to the deficiencies in comprehensively integrating various determinants of PPI within built environments, numerous PPIs were carried out in a relatively coarse and random mode, resulting in additional construction costs, land resource underuse, supply–demand mismatching, etc. The absence of the tailored tools of PPI greatly restricts the improvements of urban outdoor recreational systems and residents' well beings. Correspondingly, according to the requirements of pocket park development and current application of AI technology in urban study, this chapter explored an AI-based solution that supports the identification and prioritization of potential sites for PPI decision-making in high-density environments. The proposed solution utilized the big data of street view images (SVIs) and convolutional neural network (CNN) to construct four parallel-task models of deep learning (DL) which contributes to enhance the accuracy and efficiency of site identification and selection of PPI by

C. Zhou, *Principles and Methods of Pocket Park Planning and Design*, Urban Sustainability, https://doi.org/10.1007/978-981-96-5801-5_8

adding fine-grained assessment of 3D site condition to traditional 2D spatial analysis. We hope this approach can provide a new perspective and tool for park planning in high-density environments and promote the conversion of urban green space planning paradigm.

8.1 Integration of AI Technologies

8.1.1 Big Data Foundation

Street view images (SVIs), as an emerging data type, have already shown the capability in representing the large-scale, fine-grained, and three-dimensional conditions of the streetside spaces within the built environment, through panoramic images from pedestrian's perspective. It can not only capture the key features of the determinants within the potential sites, but also illustrate their interrelationships with the surroundings. Today, the time- and geo-tagged SVIs over the city can be freely obtained from the open-source map platforms [[5]]. Researchers can customize parameters such as vision angle, height, and other settings of street view car (SVC) according to the exact research requirements, allowing for tailored acquisition of street-front or street-side SVIs that comprehensively present the internal and external conditions of the target sites [[5]].

The popularization of SVIs worldwide provides a fine-grained and precise tool for assessing the conditions of streetside spaces over the city [[6,7]]. Existing studies demonstrated high consistency between SVI-based measurements and on-site manual surveys, further confirming the reliability of SVIs in urban studies [[8]]. The integration of SVIs with AI techniques, such as semantic segmentation models (e.g., SegNet, PSPNet, DeepLab V3) and deep learning models, can further enhance the ability to extract deep features from images, allowing for the quantification of people's perceptions in urban spaces [[9,10]] and the measurement of the socioeconomic features of the spaces, such as street vitality [[11,12]].

8.1.2 Model and Algorithm Support

Deep learning (DL) is a branch of artificial intelligence (AI) and machine learning (ML) that learns to extract features from extensive datasets by constructing and training deep neural network models [[13]]. This approach facilitates automated processing and decision-making for complex tasks which can efficiently extract features from images, audio, and text, replacing the manual feature extraction processes used in traditional machine learning [[14]]. Convolutional neural network (CNN) is a powerful deep learning architecture that effectively explore and classify hidden information in images, allowing for the extraction of deep features [[15]] and providing AI-driven solutions for urban studies. Correspondingly, CNN models have been applied in urban studies, replacing manual work in the tasks like extracting and analyzing the spatial patterns of street vendors [[16]], classifying the style and eras of street architecture [[17]], assessing the walkability of urban public spaces [[18]], and so on.

EfficientNet optimizes traditional CNN architectures by finely adjusting network width, depth, and resolution, achieving a dual leap in performance and efficiency [[19]]. It is widely applied in fields such as image classification, object detection, and segmentation. EfficientNet-B0, in particular, provides superior performance with a lower computational cost, making it well-suited for resource-constrained environments [[19]]. As the foundational version of EfficientNet, it strikes an ideal balance between accuracy and computational efficiency, making it especially suitable for processing street view image (SVI) datasets in urban planning applications [[20]].

8.1.3 Interfaces Between Technologies and Site Selection Task of PPI

The streetside spaces, where pocket parks originate from, offer a spatial interface between the application of big data technologies and site selection task of PPI. Due to their high accessibility and openness, these spaces are prioritized for PPI in planning practice, for which vast majority of existing pocket parks are located at streetside. Additionally, compared to conventional urban and community parks, pocket parks are much smaller, for which the internal and external influencing factors of PPI primarily located within a visible scope. The aforementioned two points lay a solid foundation for the application of SVIs in site selection of PPI. As the big data of SVIs can thoroughly reflect the conditions of streetside spaces of the entire city, most influencing factors of PPI can be directly and indirectly measured according to SVIs' presented information.

The perception and measurement of the 3D conditions of streetside spaces provide a working interface between the application of DL technology application and site selection task of PPI. DL technology has already shown confident performance in measuring the physical and perceptual features of the spaces according to the deep features of SVIs, which can reveal the essential factors of PPI that might be overlooked in 2D spatial analysis [[16]]. Integrating with the SVI big data, DL can efficiently map the condition of large-scale streetside spaces in a fine-grained mode. This approach not only enlarges the scope of the measurement to the citywide scale which is impossible based on manual survey, but also enhances the accuracy and stability of the measurement results compared to the experience-based judgement. Based on these two interfaces, an integral application of these two technologies can break out the temporal-spatial and precision limitations of the conventional planning methods, providing a quantified and intuitive basis for the decision-making in pocket park planning and design.

8.2 Technical Framework

8.2.1 Roadmap

To efficiently integrate the relevant technologies and data into the planning task, we aggregated four technology modules into the roadmap, including data collection, potential identification framework construction, DL model establishment, and model application (Fig. 8.1).

1. **Module 1: Data Collection**

   The Application Programming Interface (API) of open-source map platforms is used to programmatically capture SVIs of the target area for final application; training samples are randomly selected into the dataset for model training in Module 3.

2. **Module 2: Potential Identification Framework Construction**

   According to existing results of research and practice and accessible information from SVIs, the factors affecting the site selection of PPIs are classified into various categories; the assessment items of each category are specified to details and pretested for modelling.

3. **Module 3: DL Model Establishment**

   SVI samples of the dataset are randomly extracted in pairs and evaluated by experts through pairwise comparisons; the Trueskill algorithm is employed to convert the results of the various assessment categories into stable rankings and scores, which serve as the label set for samples and are input into parallel task DL models for training; multiple rounds of human-in-the-loop (HITL)

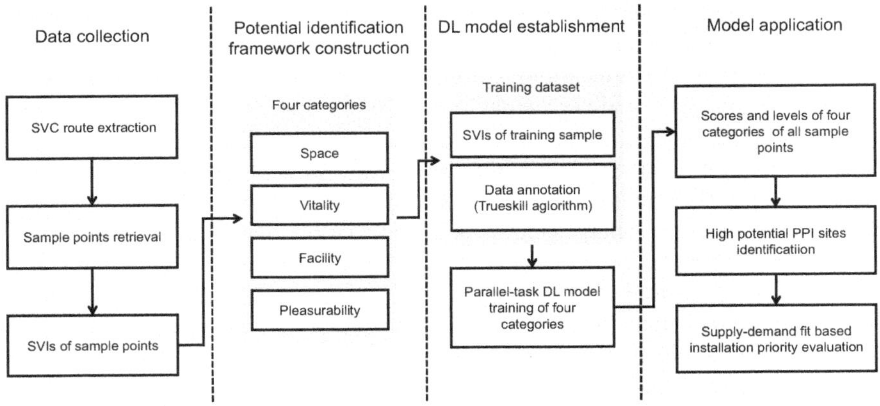

Fig. 8.1 Research roadmap

testing are carried out to progressively optimize the quality of the training set, leading to a DL model with confident performance.

4. **Module 4: Model Application**

   Trained DL models are applied to all sample points for PPI potential identification, ultimately attaining comprehensive PPI potential score and ranking of each point; Based on the supply-demand fit status of the research scope, high-potential points can be further prioritized, facilitating the final decision-making in site selection.

8.2.2 Essential Works of Each Module

The tight interconnections among the four modules raises strict requirements on the quality and precision of each module's work to ensure the reliability of the entire approach. Accordingly, the essential works of each module can be further refined as following:

1. **Module 1: Data Collection**

As SVI data serves as the foundation for all the other modules, the quality, integrity, and standardization of the SVI data can mostly ensure the efficiency and precision of the following works. Accordingly, the essential works of this module includes: (1) guaranteeing the completeness of the site condition presented in SVIs and minimizing the interference from the invalid elements (front visual obstacles like vehicles, lamp post, trees, and people); (2) adjusting the parameter in SVI collection to a degree that can guarantee the collected SVIs can cover the conditions of all the streetside spaces without any blind spot while avoid the abundance of data amount.

2. **Module 2: Potential Identification Framework Construction**

Given that the factors influencing pocket park service and installation cannot be directly extracted from the SVIs, the interpretation of the SVI information in the potential assessment of PPI is crucial to model training and the measurement accuracy. This involves three key points: (1) The site potential of PPI is determined by both internal and external influencing factors, requiring for a balanced measurement in the framework; (2) The influencing factors also cover the multiple aspects including physical, perceptual, and socio-economic, thereby asking for the representative item in the framework to ensure the measurement integrity; (3) The assessment categories within the framework should have enough independence and complementarity, requiring to avoid ambiguous definition, similar connotation, and repeated items.

3. **Module 3: DL Model Establishment**

DL model establishment is the core task of this approach, which have three essential works: (1) In order to ensure the quality and efficiency of the model training, it is essential to establish reliable, stable, and professional-based standards, processes, and tools for data annotation; (2) Building on the foundation of a reliable dataset, selecting an appropriate model that can use a relatively controllable training data to reach a relatively high precision is also crucial to guarantee the feasibility of this approach; (3) Using a proper way to training the model to guide the fine-tuning and training sample upgrading is also decisive to the efficiency and effectiveness of the model training.

4. **Module 4: Model Application**

Once the model is applied in the potential identification of PPI, there are three questions to be answered: (1) How do the assessment score of each category generated from DL model be integrated into a comprehensive potential result? (2) How do the features of the complementarity and flexibility of various categories be reflected in the quantification process of the comprehensive potential?)(3) How do the additional information be integrally measured with the SVI-based model results to comprehensively determine the priority sites for PPI?

8.3 Method

8.3.1 Data Collection

Open-source map platform was used as the data source of SVIs and the platform's API for static panorama services was employed to capture SVIs. Specifically, the sample point data with geo-located information can be preliminarily extracted from the route map of SVCs within the target study area in ArcGIS.

The image size was 600 × 480 pixels. To comprehensively capture the streetside space, a 50-m sampling interval was applied, with three sampling angles of 45°, 90°, and 135° to the SVC's driving direction in preliminary study. This method ensured continuous and integrate presentation of city streetscapes, accurately reflecting both internal and external features of streetside spaces while avoiding samples with excessive density and generating massive data. Furthermore, this SVI capture proposal also allowed for partial overlap of images from the adjacent points, facilitating diverse perspectives to present site conditions and reducing interference of the temporary obstructing factors like pedestrians, lamp posts, tree trunks, and vehicles.

To further guarantee the sufficiency of the site condition presented in SVIs, we adopted a one-way collection strategy to capture SVIs from both sides of the SVCs' route, while a two-way strategy was employed to solely retrieve SVIs from near-street side (Fig. 8.2). This action effectively mitigated the insufficient presentation of streetside condition due to the distant shooting of SVCs from the target sites.

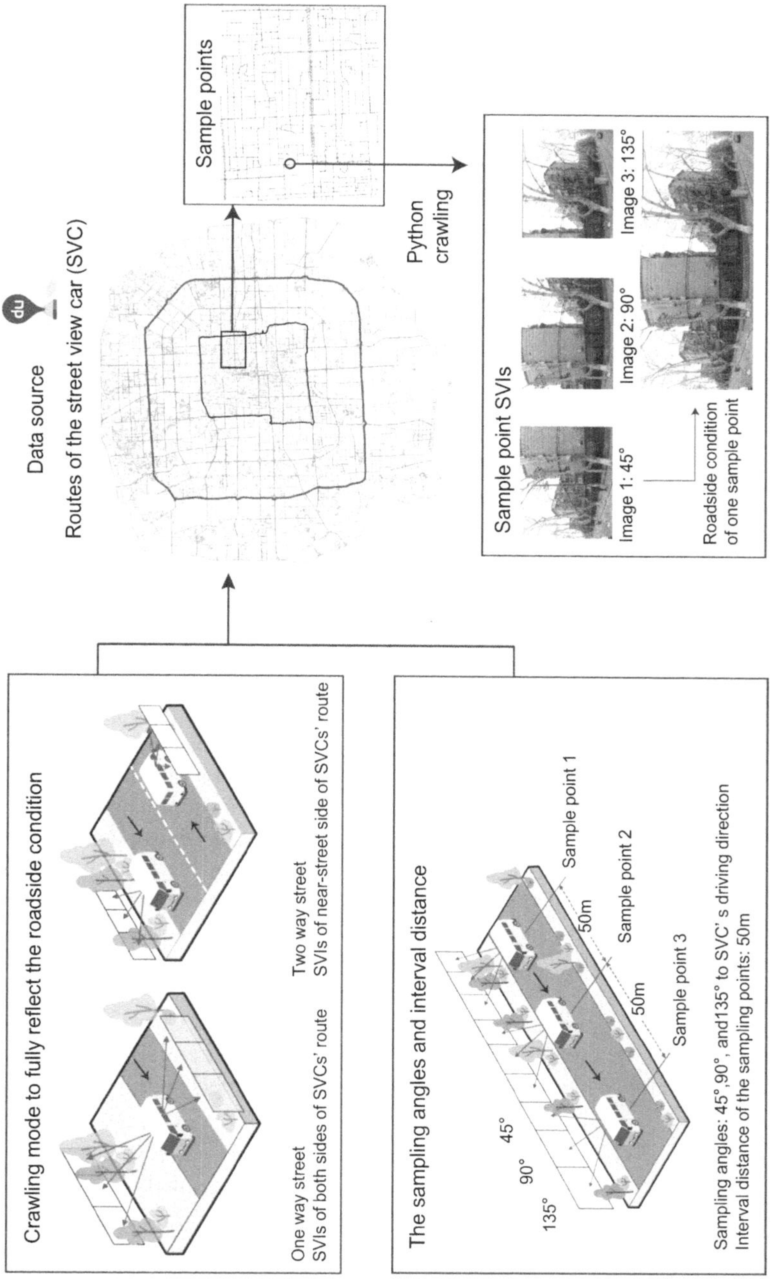

Fig. 8.2 Capturing procedure of SVIs

8.3.2 Potential Identification Framework of PPI

In planning practice, factors influencing PPI involve both internal and external conditions, covering the fields of physical, perceptual, and socioeconomic. Of these factors, internal factors relate to the site's inherent attributes, such as site size, shape, enclosure degree, pavement, greenery, and recreational facilities. These primarily provide the basic physical conditions for park installation and shape visitors' sense of place. External factors refer to the surrounding characteristics of the site, such as the type of surrounding facilities, potential user density, land use and value, and the aesthetics of surrounding buildings. Collectively, these factors offer external support or inhibition for PPI from the aspects of physical, perceptual, and socioeconomic.

The frequently-raised question is how can the status of site external factors be assessed based on SVIs, since the visual field of SVIs is strongly limited. According to a careful examination of the characteristics of the information shown by SVIs, we found that due to the short-radius (approximately 300 m or a 5-min walking distance) service mode [[21,22]]adopted by pocket parks, the site surrounding information shown by SVIs demonstrated high representativeness of pocket parks' entire service scope. For instance, the height and density of surrounding buildings, as reflected in SVIs, can effectively indicate the density level of pocket parks' potential users distributed around [[18,23]]. Although this is not fully precise, it can basically meet the modelling requirements of this study. Accordingly, these features were also integrated into the identification framework in order to compensate for the deficiencies of SVIs-based assessment methods.

Additionally, to facilitate the subsequent work on training sample annotation and DL modelling, the framework construction needs to ensure that the items of each assessment category are distinct, thereby avoiding the repeat or correlated assessment which would weaken the effectiveness of the identification result. Meanwhile, existing studies proved that different factors supporting PPI can compensate for each other under specific conditions. For example, surrounding facilities such as retail shops and public toilets can make up for the shortcomings of the facility absence in pocket parks which is resulted by space limitation [[24,25]]. Accordingly, it is also essential to reserve flexibility in the construction of the identification framework, enabling the factors of different categories to complement with each other in PPI potential assessment.

According to existing studies, factor framework presented in Chapter 3, and the specific requirements of our task, we constructed the PPI potential identification framework based on four aspects: space, vitality, facility, and pleasurability (Fig. 8.3).

1. **Space**

Different from traditional parks whose size are rigidly controlled, pocket parks' size and form are more flexible, leaving them with higher adaptability to the built environment. However, as different daily recreational activities raise specific requirements on pocket parks' size, shape, and boundaries, space suitability of potential sites is considered as a prerequisite for PPI [[26]]. For example, Ashihara (1970) argued

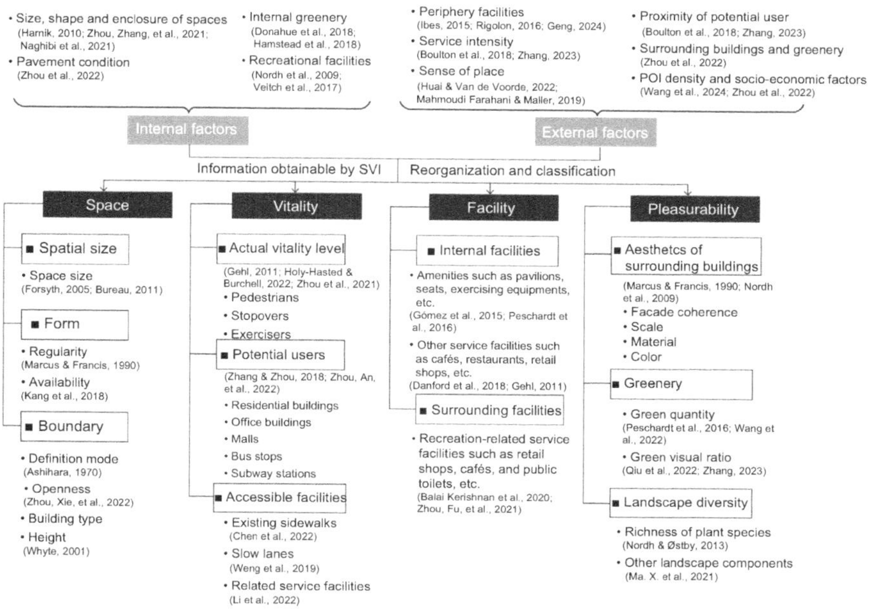

Fig. 8.3 PPI potential identification framework

that an appropriate aspect ratio of small public spaces is crucial for enhancing users' experience [[27]]. In Hong Kong, the *Public Open Space in Private Development Design and Management Guidelines* (2011) mandated that the size of the pocket spaces should be larger than 100 m^2 while their aspect ratio should be less than three to ensure an effective utilization [[28]]. As discussed in Subsections 2.2.4 and 3.1.3, boundary openness and functions of boundary buildings also significantly influence the vibrancy and service flexibility of pocket parks. For instance, longer open boundary might significantly promote the park visits, thereby guaranteeing parks' service effectiveness.

Accordingly, the key points for assessing the space suitability were categorized into: (1) Spatial size—the size of the available space including various streetside open spaces; (2) Spatial form—the regularity and availability of streetside open space; and (3) Boundary—the definition mode, openness, and building type and height of the open space boundary.

2. **Vitality**

Site vitality indirectly indicates site's accessibility and proximity to potential users, which is considered as the core value of pocket parks' service [[1,29]]. Since SVIs only demonstrate the instantaneous status of streetside sites, to reduce the bias and integrally assess the site vitality, more and more efforts are being made to depict the potential vitality based on the information of SVIs. For example, existing studies have already managed to use SVIs to indirectly evaluated the accessibility of streetside spaces according to the activity intensity, barriers, sidewalk integrity, and walking-related infrastructure [[18]].

Accordingly, based on accessible SVI information, the site vitality was delineated into three points: (1) Actual vitality level—the crowd size of existing pedestrians, stopovers, and exercisers shown by the SVIs; (2) Potential users—the size and density of residential buildings, office buildings, malls, bus stops, and subway stations around the site; and (3) Accessibility facilities—existing sidewalks, slow lanes, and related service facilities.

3. **Facility**

Well-structured and high-quality service facilities can inspire higher willingness to visit parks [[30]]. As pocket parks cannot be equipped comprehensively like larger parks, surrounding facilities such as retail shops, cafes, and public toilets can compensate for the corresponding facility absence in the parks [[24,31]]. As Subsection 3.2.1 mentioned, this feature has already been confirmed by related studies, which indicate that the density and diversity of surrounding facilities can significantly increase the visits of pocket parks [[32]]. Moreover, existing recreational facilities within the target sites—such as tables, chairs, fitness equipment, service kiosks, public toilets, and bike racks—also contribute to a more favorable condition for PPI, which can greatly save the construction cost [[33,34]] and provide an innate base of place familiarity for surrounding residents.

Therefore, the identification of facility conditions was characterized by the following aspects: (1) Internal facilities—amenities and other recreational facilities within the target sites; and (2) Surrounding facilities—recreation-related service facilities, such as retail shops, cafes, and public toilets, around the target sites.

4. **Pleasurability**

Pleasurability refers to how individuals perceive the park space, which directly influences its attractiveness and indirectly determines the frequency of park usage [[35,36]]. Existing studies on public spaces suggested that pleasurability encompasses the space attributes of diversity, vibrancy, enjoyment, and walkability [[18]], which is mainly affected by the site internal factors such as greenery and landscape diversity and external factors like the aesthetics of surrounding buildings [[37–41]].

Consequently, pleasurability was identified through three aspects: (1) Aesthetics of surrounding buildings—facade coherence, scale, material, and colour; (2) Existing green spaces—green quantity and green visual ratio; and (3) Landscape diversity—the richness of plant species and other landscape components.

8.4 DL Model Establishment

1. **Model Structure and Mechanism**

Corresponding to the four assessment categories in the potential identification framework of PPI, we constructed four parallel-task DL models (Fig. 8.4) adopting convolutional neural networks (CNNs). Compared to the general classification tasks, the CNN-based non-linear regression task excels in automatically learning and extracting deep features from large volumes of images, yielding continuous percentile scores with higher precision, interpretability, and uncertainty capture [[13]]. Moreover, CNN owns a virtue that can achieve a better precision with a relatively controllable training workload, making it widely applied in the information recognition and classification of abundant images.

Each task model consists of three parallel backbone networks to extract features from the input images, and the subsequent neural network layers that perform comprehensive analysis on the feature extraction results of the three images for each sample point. The pre-training backbone network utilizes the highly adaptable lightweight convolutional neural network EfficientNet-B0, which employs a compound scaling method, with superior scalability to ensure excellent performance in feature extraction tasks across various applications [[19]]. By handling the SVIs from three different angles through the three backbone networks, the model can extract unique features under each angle with specific trained parameters. The convolution, pooling, and activation in the backbone networks enable the parallel mining and synthesis of features from the three images at different angles, generating a

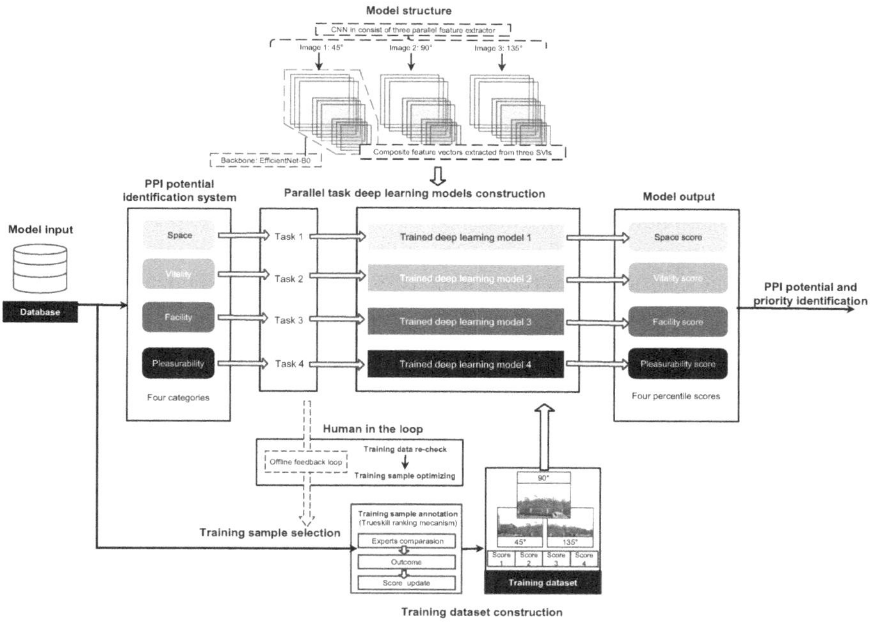

Fig. 8.4 Establishment of Parallel-task deep learning model

composite feature vector. Finally, the model used a dense layer to quantify the fitness of the composite feature vector and output the percentile score.

2. **Training Dataset Construction**

According to the existing DL regression tasks, a pre-trained sample size of 500 is recommended to effectively balance the model performance and training cost. To ensure the representativeness of the training dataset, the proposal initially involved 10,000 randomly-selected samples from the data set with an even and dispersed distribution over the study area. Consequently, 500 samples were randomly selected from the 10,000 preliminarily samples which were manually validated, compared, and replaced based on the potential identification of four categories to ensure that the training samples showed adequate representativeness of the different associated patterns of the four categories exist at different levels.

To construct a relatively standardized training dataset, we adopted the Trueskill algorithm to ensure the stability and reliability of the scoring across the four categories. Trueskill is a Bayesian-based rating system that generates ranking scores through pairwise voting. It was originally developed for gaming and competitive sports. The system accumulates results from pairwise competitions to create ranked scores for winners and losers, which are iteratively updated after every two-player contest and can effectively convert binary comparison results into continuous ranking scores [[42]]. Compared to classic Elo rating systems, the Trueskill-based pairwise voting method is more flexible and scalable for handling complex scenes [[43]], and it has been proved to be a mature data annotation technique in existing urban studies that integrate DL and SVIs, such as the Place Pulse 1.0 [[44]] and Place Pulse 2.0 datasets [[45]]. Additionally, Trueskill shows great value in transforming subjective ranking into stable pairwise comparisons, thereby addressing the inherent uncertainty in the identification process.

The Trueskill algorithm relies on accurate win-lose outcomes, while the assessment of PPI potential site greatly depends on domain-focused expert judgment to ensure comparison accuracy. Therefore, it is suggested that all participants involved in the data annotation should be professionals in urban planning and landscape architecture to guarantee the judgement accuracy and validity. Additionally, each streetside sample site should be represented by three SVIs from different angles to comprehensively show the internal and external conditions, reducing the likelihood of bias in comparisons. Since the result is measured by the level of each category with three grading scores rather than quantitative units or indicators, this annotation and assessment method fully meets the precision requirements of the study.

To ensure the operability of the voting process, we developed a web program for pairwise comparison of SVIs based on Trueskill algorithm (Fig. 8.5). According to the proposed potential identification framework, the program included four voting modules: "Space", "Vitality", "Facility", and "Pleasurability". During comparison, the web backend randomly selected a pair of SVIs from the training dataset for pairwise voting at front end. Voting was a contest, and the sample whose SVIs got selected was the winner. It did not allow a draw and the participants were only provided with the sample images, without access to any other information like images' geolocations.

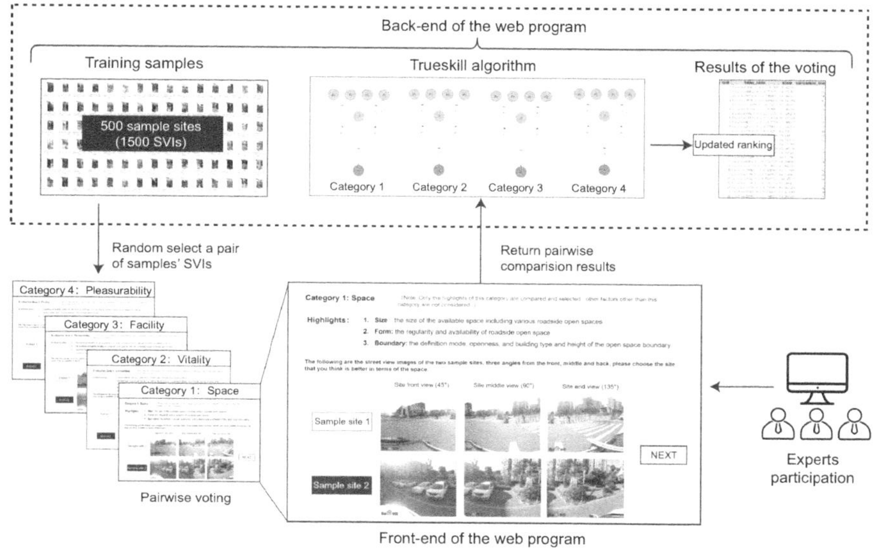

Fig. 8.5 Pairwise comparison program and Trueskill-based scoring mechanism

Based on the results of each round of pairwise voting, the program iteratively updated the scores of all the training samples in the four categories before the completion of the pair-wise voting task. Generally, a Trueskill algorithm requires 22–32 votes per sample to achieve 75% internal consistency in the ranking results [[46]].

Additionally, a human–machine interaction approach—'human-in-the-loop (HITL)' was employed to guarantee the applicability and precision of the model, as it can incorporate human knowledge and wisdom into the learning process, which improves the model's performance with minimum labor and time [[47]]. HITL can be implemented at various levels, depending on who controls the workflow. In this study, an active learning (AL) approach was adopted [[48]], where the system remained in control and employed humans as oracles to annotate data. The objective was to minimize human annotation while maximizing the model's performance. AL is particularly effective in the scenarios that there are abundant unlabeled data, whereas annotation tasks are costly or time-consuming. Once there is a great discrepancy between the model's output and the user's expected results, new samples are selected, annotated, and supplemented, or existing imprecise annotations are corrected in the iterative feedback loops. In this iterative process, the model's performance is sustainedly enhanced [[49,50]].

In our study, experts randomly selected 200 unannotated samples from the SVI dataset and generated rankings for each sample based on their percentile scores. Samples whose model-generated scores consisted (within ± 10%) with the expert-generated scores were included in the maturity calculation. The samples with inconsistent scores were analyzed for the corresponding tunning of the training data set, based on which more representative labelled samples were supplemented while the labelled samples that deviated significantly from the realistic data were either removed or re-annotated [[50,51]].

8.5 Model Application

1. PPI Potential Identification

Given that the potential of PPI was jointly determined by the assessment results of the four categories of the DL models, a two-phase grading method was carried out for potential identification (Fig. 8.6). In the first phase, the grading scores for all the categories were derived based on the results of K-means binning. K-means binning is employed in the data post-processing for discretizing the continuous score results, distinct from its clustering use in unsupervised learning. Its advantage lies in dimensionality reduction, noise reduction, and enhanced model stability. Compared to other binning methods, K-means binning offers greater flexibility, interpretability, and suitability for various data types [[52]].

Four parallel-task DL models were employed for generating scores of all sample points. After standardization, k-means binning was utilized to categorize scores of each category into three bins based on cluster centroids, assigning grading scores

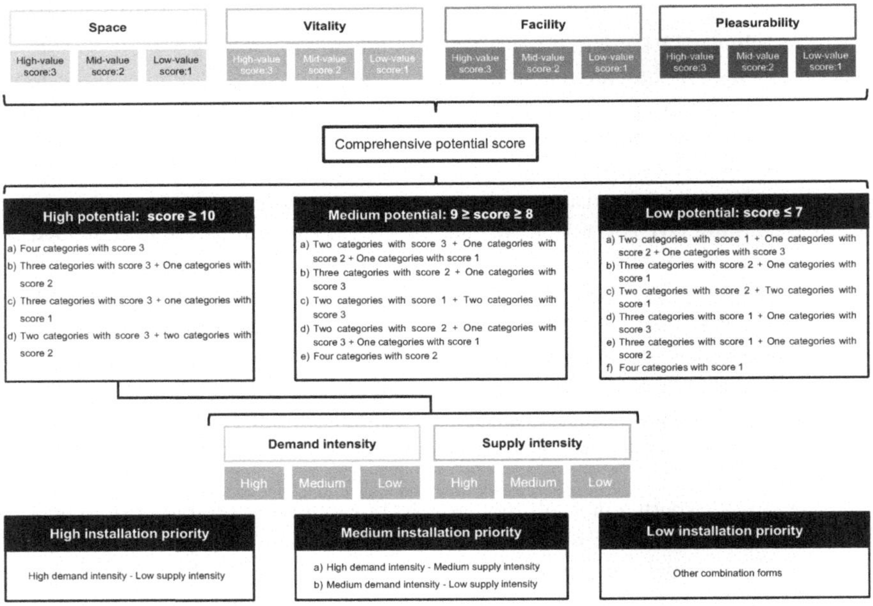

Fig. 8.6 "Potential-Priority" identification framework

of 3, 2, and 1 for high, medium, and low clusters, respectively. The comprehensive potential score for each sample point was then calculated by summing the grading scores of four categories. According to the comprehensive potential score, sample points can be classified into three classes with high (10–12 points), medium (8–9 points), and low (4–8 points) potential of PPI.

2. **Installation Priority Assessment**

Building on the assessment results of the PPI potential focusing on the site conditions, the priority sites of PPI can be further determined through a consideration of the status of supply–demand fit in a broader scale, as mentioned in Chapters 5 and 7. The combination of these two approaches offers an opportunity of the cross-scale examination on PPI site identification, further enhancing the precision and reliability of the results. Accordingly, we proposed a two-phase approach focusing on the identification of potential and priority, respectively, to finally derive the sites' prioritization for PPI, especially for the sites with high potential.

Different from the approaches applied in Chapter 7, the supply–demand relationship was directly assessed focusing on the intensity comparison of the existing park service provision and demand, without consideration of the existing supply conditions (e.g. existing recreation-related facilities, slow traffic network, and so on), as most factors of the supply conditions were already assessed in the PPI potential identification based on SVIs and DL models. After standardizing the existing supply and demand intensity values, they were clustered into three levels: high, medium, and low. By associating the supply and demand levels in space, the supply–demand relationship of different areas can be quantified and visualized for prioritizing PPI. Of the points with high potential of PPI, those located in "high demand—low supply" areas were annotated as high-priority sites for PPI; those located in the "high demand - medium supply" and "medium demand - low supply" areas were annotated as medium-priority sites for PPI; while those located in the other areas were annotated low-priority sites for PPI (Fig. 8.6).

8.6 An Application Case

8.6.1 Study Area

Beijing, the capital of China, serves as the country's economic, cultural, and political center. In 2022, the city's total population was 21.86 million, with an average density of 14,800 people/km^2 in built-up areas [[53]]. In some areas of the old city or CBD, the density exceeded 30,000 people/km^2, exemplifying a typical high-density environment.

Under the pursuit of refined renewal and construction in the past decade, the Ministry of Housing and Urban–Rural Development of China released the *Guidelines for Pocket Park Construction* in 2024 [[54]]. It provided unified planning guidance for the spatial distribution, design, and maintenance of pocket parks in order to promote their nationwide development. In the same year, the *2024 Action Plan for Improving the Quality of Beijing's Urban Slow Traffic System* [[55]] was released and implemented, advocating for new installation of pocket parks through conversion of the existing land use and utilization of the interstice or vacant lands. This initiative highlighted the vital role of pocket parks in improving urban's social and environmental sustainability.

Guided by the long-term planning, Beijing's green space development has made significant progress and its per capita area of parkland reached to 16.9 m^2 in 2023, exceeding the national average level of 15.3 m^2[[56,57]]. However, notable disparities exist in the distribution of public parks between the urban center and peripheral areas due to the neglect of the green space development in the early urbanization process. The continuous growth of urban population in the central districts, coupled with limited parklands, exacerbates the mismatch of the recreational resource supply and demand. Thus, considering the scarcity of land resources within the built environment, pocket parks are relied as the main tool to promote the increase of outdoor recreational opportunities in Beijing, through which the supply–demand contradiction is expected to be mitigated.

We selected the area within the Fourth Ring Road of Beijing as the study area, mainly for three reasons: (1) As a densely populated area with a high level of urbanization, the findings can be highly relevant and applicable for other similar high-density urban areas; (2) There is a significant discrepancy in the distribution of recreational resources in the study area, necessitating a allocational rebalance through pocket park planning; (3) Beijing's built environment is diverse and the SVI data is of high coverage and timeliness, providing a substantial dataset for model training and application in PPI identification.

To facilitate cross comparison between different featured zones, we further divided the area into: (1) the old city that is enclosed by the Second Ring Road, serving as the core functional district of the capital whose per capita parkland area was only 7 m^2 (2022) [[58]], and (2) the expanded zone that is located between the Second and Fourth Ring Road, functioning as the expanded support district of the old city whose per capita parkland area reaches to 16 m^2 (2022) [[58]](Fig. 8.7).

We utilized Baidu Maps, one of the most widely used mapping platforms in China, as the source of SVI data. After data cleaning and filtering, a total of 122,309 sampling points and 366,927 SVIs were obtained within the study area. They were primarily captured in December 2022 when the updated conditions of streetside spaces can be accurately reflected. 500 points with 1500 SVIs were randomly selected as the training samples. After the manual annotation of 12 experts, the average number of votings for the four categories was 30.412, 29.596, 30.052, and 29.660, respectively, meeting the requirement for the stable output from the Trueskill algorithm ($\geq$12 votes).

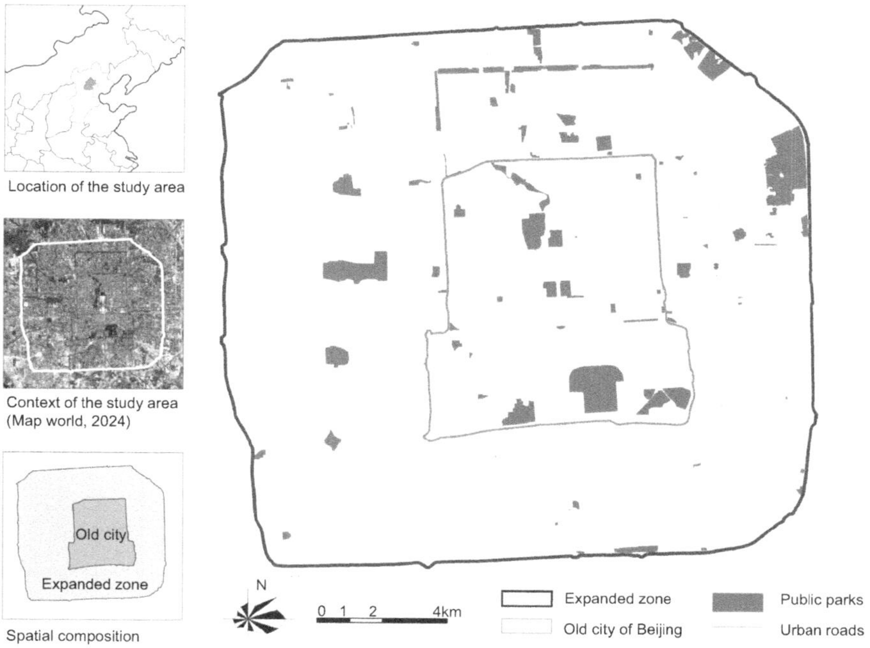

Fig. 8.7 Study area

8.6.2 *Results of DL Models Application*

According to the results of DL model training, the values of Mean Squared Error (MSE), Root Mean Square Error (RMSE), and Mean Absolute Error (MAE) indicated that all the four models demonstrated high accuracy (MSE < 0.06, RMSE < 0.225, MAE < 0.19). After 4–5 rounds of Human-in-the-Loop (HITL) tuning, each model achieved a maturity level exceeding the conventional standard (maturity > 0.75), signifying robust performance and reliability.

The binning results from the four model outputs (Fig. 8.8) indicated that the proportion of the high-score samples in all the categories was the lowest. Notably, there were differences in the proportion of high-score samples between the two subzones within the study areas. For example, the proportion of the high-score samples of space category (19%) in the old city is lower than that in the expansion area (25%). Conversely, the old city had a significantly higher proportion of high-score samples for facility category (21%) compared to the expanded zone (13%). Additionally, the old city also showed higher proportions of the high-score samples of both vitality and pleasurability categories, compared to the expanded zone. This suggests that the expanded zone owns better spatial conditions for pocket park installations, while the old city excels in the conditions of facility, vitality, and pleasurability. This features was confirmed again by the results of the Getis-Ord Gi* hotspot analysis,

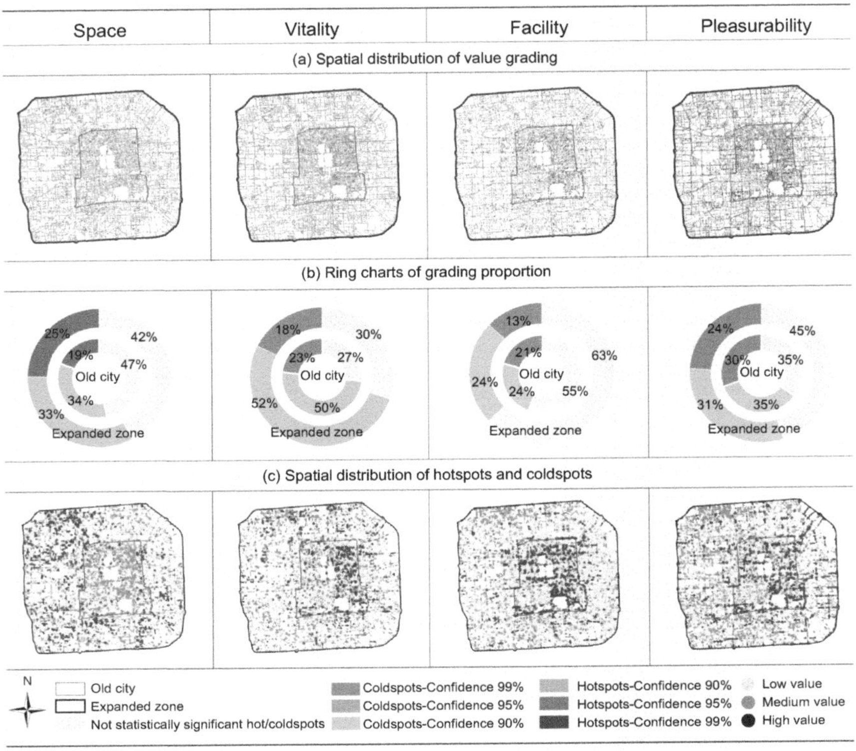

Fig. 8.8 Evaluation results of four potential identification categories

aiming to examine the spatial autocorrelation characteristics of the assessment results of the four categories.

8.6.3 Distribution of Potential Sites

After the two-phase grading process, the grading scores of the sample points were distributed as follows (Fig. 8.9): high potential (6.08%), medium potential (31.90%), and low potential (62.02%). Notably, the proportion of high potential sites in the old city (8%) was higher than that in the expanded zone (5%), indicating that the conditions for pocket park development in the old city are superior to those in the expanded zone.

Further analysis of the spatial autocorrelation characteristics of the 7,432 high potential points (Fig. 8.10) revealed that the high potential sites clustered in the northeast, southeast, and certain western areas of the old city, denoting relatively favorable conditions for pocket park development in these areas. In contrast, the high

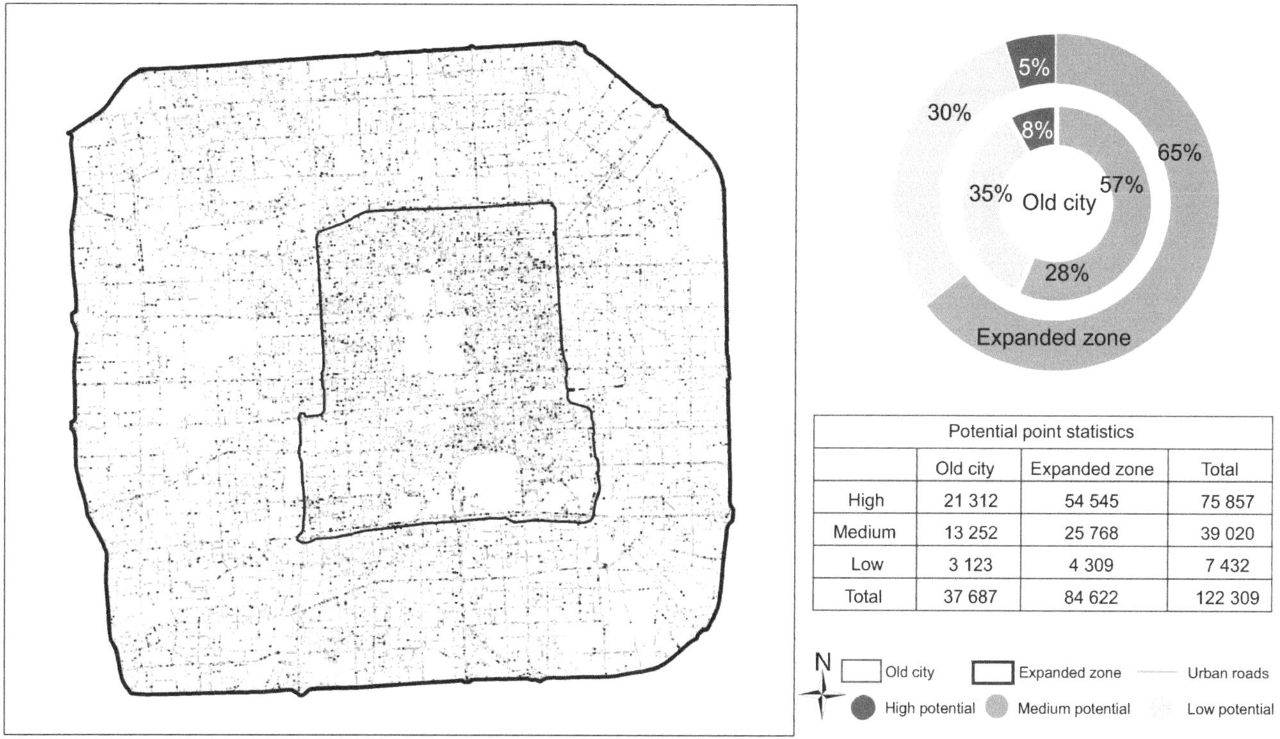

Potential point statistics			
	Old city	Expanded zone	Total
High	21 312	54 545	75 857
Medium	13 252	25 768	39 020
Low	3 123	4 309	7 432
Total	37 687	84 622	122 309

Fig. 8.9 Distribution of potential levels

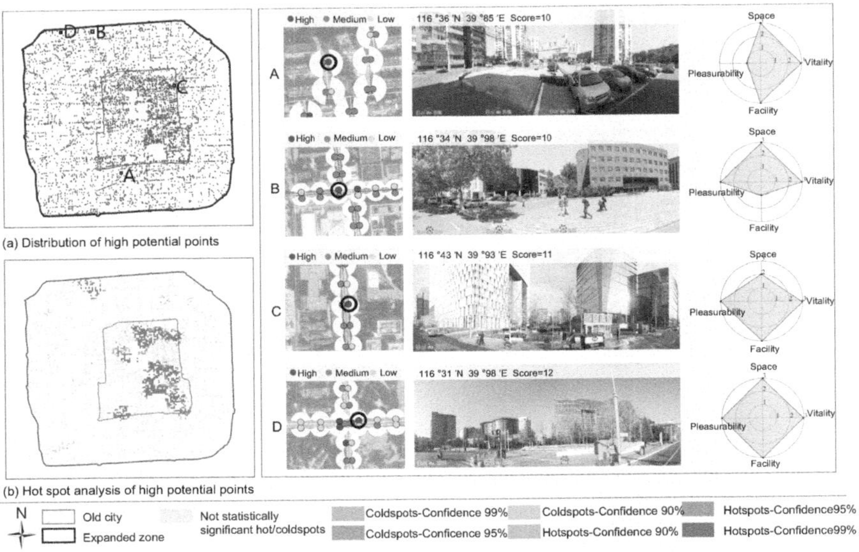

Fig. 8.10 Distribution of high potential sites

potential points only clustered locally in the northwest of the expanded zone, with no equivalent cluster elsewhere, indicating a more even and homogeneous conditions for pocket park development at this zone.

8.6.4 *Prioritization of Potential Sites*

The "Potential-Priority" identification results (Figs. 8.11 and 8.12) indicated that high-priority sites of PPI account for 33.19% (a total of 2,467 points). There were notable differences in the proportion of high-priority points between the old city (39%) and the expanded zone (29%). Considering the area of the old city is much less than the expanded zone, the density of the high-priority points of PPI is much higher within the old city, which further confirms the severe scarcity of outdoor recreational spaces and the urgency of pocket park development in the old city.

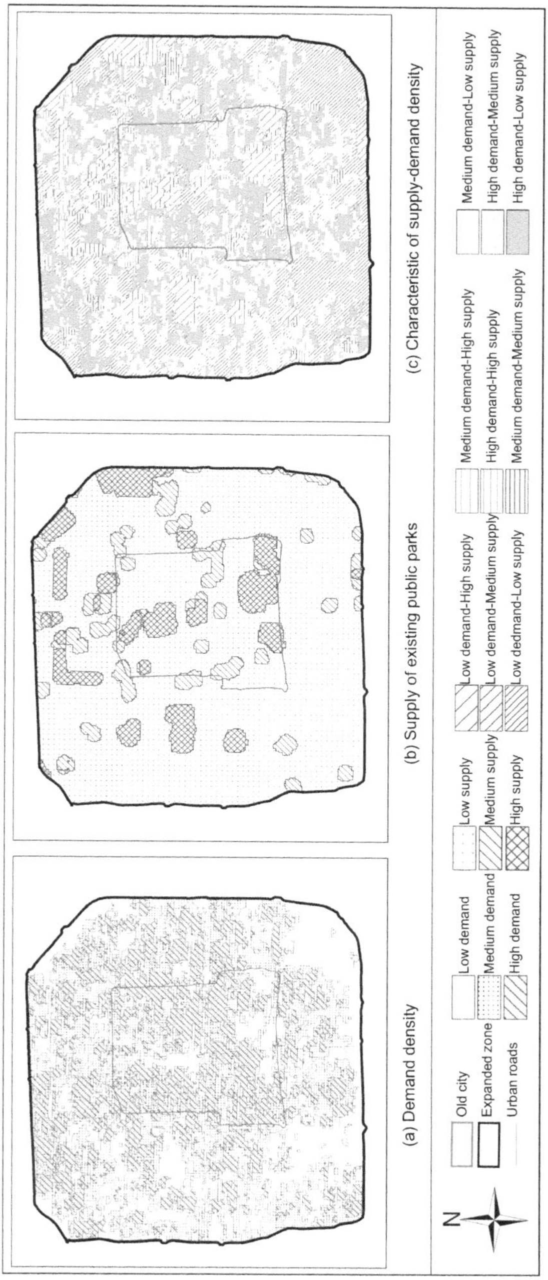

Fig. 8.11 Supply–demand density of recreational resources

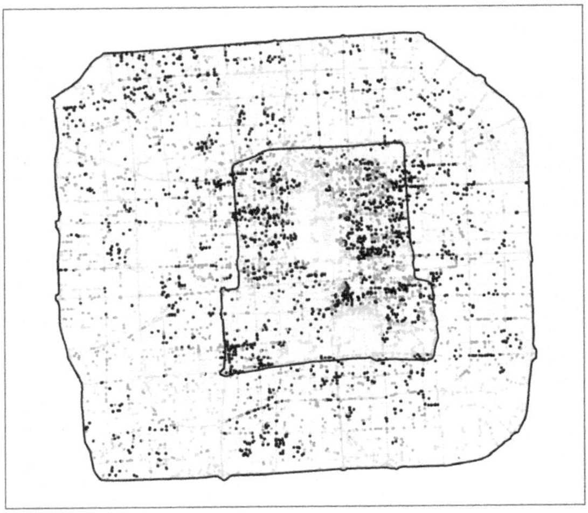

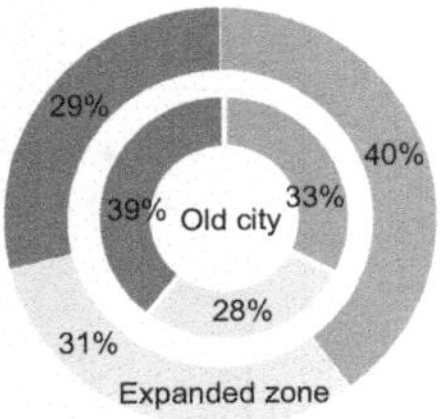

Installation priority of high potential points			
	Old city	Expanded zone	Total
High	1 018	1 179	2 737
Medium	875	1 353	2 228
Low	1 230	1 237	2 467
Total	3 123	4 309	7 432

Fig. 8.12 Priority of potential PPI sites

8.7 Enlightenments and Conclusions

8.7.1 Method Innovations

By fully utilizing the Trueskill algorithm, EfficientNet-B0 backbone network, and HITL active learning method, we developed four parallel-task DL models with controllable time and labor costs, providing an innovative perspective and approach for pocket park planning. This parallel task modeling approach, in conjunction with SVIs, can effectively classify and assess multiple factors of the built environment in a large scale, breaking through the spatial and temporal limitations of traditional surveys. The flexibility and compatibility of this approach framework allows for modifying the assessment categories based on specific needs, making it applicable in the identification of other sites.

Compared to previous research and practices in pocket park site selection, the four parallel-task DL models developed by this study effectively measured the internal and external influencing factors of PPI from a fine-grained mode and three-dimensional perspective, compensating for the inherent limitations led by the coarse granularity and two-dimensional analyses of traditional methods [[18]]. Additionally, the framework of four parallel-task DL models and the method adopted in the PPI potential calculation also enable the mutual compensation of the various positive factors related to PPI, facilitating the formulation of the tailored PPI proposal according to the site conditions.

To minimize the uncertainty and bias led by the manual labeling of the abundant samples, we employed a pairwise comparison method. Two approaches were made to ensure the credibility and accuracy of the comparison results. First, we standardized the parameters such as angles, distances, and field of view for each SVI to ensure the comparability of the site condition of each sample point. Second, we employed multiple rounds of pairwise comparisons to yield stable potential rankings through the Trueskill algorithm. Although the generated results are not measured by quantitative units as traditional measurements, they can fully meet the precision requirements of this study which only requires to annotate the level of each assessment category with three grading scores.

Moreover, the proposed "Potential-Priority" two-phase identification method created a channel to integrate the assessment of 3D site conditions with the analysis of supply–demand relationship of recreational resources focusing on the 2D spatial analysis, by which these two methods can work complimentarily in PPI site identification, thereby generating more comprehensive and valuable results to support the pocket park planning.

8.7.2 Practical Application for Planning and Design

PPI results generated by DL models can be employed for systematically classifying the potential resources of pocket park development in the built-up areas. On this base, an additional analysis of demand–supply relationship of recreational resources can provide a basis for determining the constructional sequence of pocket parks in the planning area. Furthermore, the identification results of potential sites at the urban scale can directly reflect the spatial distribution of the future development opportunities of pocket parks, establishing the foundation for constructing the service system construction, group, or sequence of pocket parks at the city scale.

The multidimensionality and flexibility of the PPI potential identification framework are helpful to identify potential sites with different characteristics, such as abandoned, marginalized, and, inefficient sites within the cities, as well as the sites with existing greening that are suitable to be converted into pocket parks. Although this framework is difficult to distinguish existing parks from the high potential sites, further analysis of supply–demand relationship can effectively compensate for this deficiency by prioritizing the potential sites of PPI. The score structure of the four categories can also serve as a basis for formulating design guidelines of pocket parks with different site conditions. For example, for the potential sites with low score of space and high scores of vitality, facility, and pleasurability, the focus in the subsequent design of pocket park should be set on integrally utilizing the surrounding facilities and landscape to make up for the lack of space in the site. Meanwhile, vertical greening, movable greening, and other ground-greening substitutional approaches could be used to reserve more ground space for recreational activity, aiming to maximumly meet the demands of surrounding residents. Similarly, other specific design and implementation strategies for PPI can also be derived from different score structures.

In this case study, we provided a quantitative analysis tool for citywide identification of PPI sites. However, besides quantitative analysis, the final decision of

PPI also needs to involve the qualificative analysis, especially on the aspects of the group preferences, land acquisition opportunities, and fund raising. This tool can significantly reduce the scope and sample size of qualitative analysis and minimize the misjudgments in decision-making. Furthermore, the proposed comparison and prioritization framework also reserves the interfaces with the subsequent qualitative analysis, facilitating the formulation of the more comprehensive plans through both qualitative and quantitative approaches.

8.7.3 Future Work and Explorations

Given the reliance on SVI big data, the completeness of SVC route restricts the spatial scope of analysis. For instance, the potential sites, that are enclosed by buildings or located in the inner blocks, may be omitted by this tool. Even if these sites may not be the optimal choices for PPI, they still provide valuable opportunities of adding recreational spaces for residents in high-density environments, where recreational spaces might be extremely scarce. This limitation can be addressed by further optimization of the accuracy and breadth of the data sources.

Furthermore, the PPI potential identified by this tool is based on the quantification of the visual information extracted from SVIs. However, there are still physical factors and non-physical factors (such as cultural, social, and policy influences) that are difficult to be obtained and quantified through SVIs. They also play crucial roles in the final decision-making of PPI. Therefore, although this method can integrate various influencing factors of PPI into site potential measurement and provide a useful basis for planning and design, it cannot entirely replace the decision-making process that requires for the participation and negotiation of different stakeholders as well as the additional in-depth qualitative analyses. Correspondingly, a smarter decision-making assistance tool is expected to be established in the future with a joint use of diverse assessment techniques and data sources.

References

1. Zhou C, Zhang Y, Fu L, Xue Y, Wang Z (2021) Assessing mini-park installation priority for regreening planning in densely populated cities. Sustain Cities Soc 67:102716. https://doi.org/10.1016/j.scs.2021.102716
2. Fatiah AA, Ponrahono Z, Zakariya K (2021) Quality of designs and features of small urban green spaces in Petaling Jaya Town, Malaysia. Planning Malaysia, 19. https://doi.org/10.21837/pm.v19i15.931
3. Dong J, Guo R, Guo F, Guo X, Zhang Z (2023) Pocket parks-a systematic literature review. Environ Res Lett 18(8):083003. https://doi.org/10.1088/1748-9326/ace7e2
4. Chen Y, Liu X, Gao W, Wang RY, Li Y, Tu W (2018) Emerging social media data on measuring urban park use. Urban For & Urban Green 31:130–141. https://doi.org/10.1016/j.ufug.2018.02.005

5. Anguelov D, Dulong C, Filip D, Frueh C, Lafon S, Lyon R, Ogale A, Vincent L, Weaver J (2010) Google street view: Capturing the world at street level. Computer 43(6):32–38. Computer. https://doi.org/10.1109/MC.2010.170
6. Biljecki F, Ito K (2021) Street view imagery in urban analytics and GIS: A review. Landsc Urban Plan 215:104217. https://doi.org/10.1016/j.landurbplan.2021.104217
7. Kelly CM, Wilson JS, Baker EA, Miller DK, Schootman M (2013) Using Google street view to audit the built environment: Inter-rater reliability results. Ann Behav Med 45(1):108–112. https://doi.org/10.1007/s12160-012-9419-9
8. Griew P, Hillsdon M, Foster C, Coombes E, Jones A, Wilkinson P (2013) Developing and testing a street audit tool using Google street view to measure environmental supportiveness for physical activity. Int J Behav Nutr Phys Act 10(1):103. https://doi.org/10.1186/1479-5868-10-103
9. Imadegawa Y, Oki T, Ogawa Y, Zhao C (2023) Predicting impression evaluation of building exterior appearance using street image big data and deep learning. IEEE International Conference on Big Data (BigData) 2023:1419–1428. https://doi.org/10.1109/BigData59044.2023.10386316
10. Xu X, Qiu W, Li W, Liu X, Zhang Z, Li X, Luo D (2022) Associations between street-view perceptions and housing prices: Subjective vs. objective measures using computer vision and machine learning techniques. Remote Sensing 14(4), Article 4. https://doi.org/10.3390/rs14040891
11. Li J, Tang P (2023) Multisource analysis of big data on street vitality using GIS mapping and deep learning: A case study of Ding Shu, China. 565–574. https://doi.org/10.52842/conf.caadria.2023.1.565
12. Wang M, Vermeulen F (2021) Life between buildings from a street view image: What do big data analytics reveal about neighbourhood organisational vitality? Urban Studies 58(15):3118–3139. https://doi.org/10.1177/0042098020957198
13. Zhou ZH (2021). Machine Learning (S. Liu, Trans.; 1st ed. 2021 edition). Springer.
14. Khan S, Naseer M, Hayat M, Zamir SW, Khan FS, Shah M (2022) Transformers in vision: A survey. ACM Comput Surv 54(10s):1–41. https://doi.org/10.1145/3505244
15. Ibrahim MR, Haworth J, Cheng T (2020) Understanding cities with machine eyes: A review of deep computer vision in urban analytics. Cities 96:102481. https://doi.org/10.1016/j.cities.2019.102481
16. Liu Y, Liu Y (2022) Detecting the city-scale spatial pattern of the urban informal sector by using the street view images: A street vendor massive investigation case. Cities 131:103959. https://doi.org/10.1016/j.cities.2022.103959
17. Sun M, Zhang F, Duarte F, Ratti C (2022) Understanding architecture age and style through deep learning. Cities 128:103787. https://doi.org/10.1016/j.cities.2022.103787
18. Li Y, Yabuki N, Fukuda T (2022) Measuring visual walkability perception using panoramic street view images, virtual reality, and deep learning. , 86, 104140. https://doi.org/10.1016/j.scs.2022.104140
19. Tan M, Le Q (2019) EfficientNet: Rethinking model scaling for convolutional neural networks. Proceedings of the 36th International Conference on Machine Learning, 6105–6114. https://proceedings.mlr.press/v97/tan19a.html
20. Ramanna S, Sengoz C, Kehler S, Pham D (2021) Near real-time map building with multi-class image set labeling and classification of road conditions using convolutional neural networks. Appl Artif Intell 35(11):803–833. https://doi.org/10.1080/08839514.2021.1935590
21. The Trust for Public Land. (2020). Pocket park toolkit. The trust for public land. https://www.tpl.org/resource/pocket-park-toolkit
22. Ministry of Housing and Urban-Rural Development, PRC. (2022). *Instruction on promoting the construction of "Pocket Parks."*
23. Chen L, Lu Y, Ye Y, Xiao Y, Yang L (2022) Examining the association between the built environment and pedestrian volume using street view images. Cities 127:103734. https://doi.org/10.1016/j.cities.2022.103734

24. Gómez E, Baur JWR, Hill E, Georgiev S (2015) Urban parks and psychological sense of community. J Leis Res 47(3):388–398. https://doi.org/10.1080/00222216.2015.11950367
25. Zhou C, An Y, Zhao J, Xue Y, Fu L (2022) How do mini-parks serve in groups? A visit analysis of mini-park groups in the neighbourhoods of Nanjing. Cities 129:103804. https://doi.org/10.1016/j.cities.2022.103804
26. SPUR (2009) Secrets of San Francisco: Where to find our city's POPOS—Privately owned public open spaces. Francisco Planning + Urban Research Association (SPUR).
27. Ashihara Y, 1918–2003 (1970) Exterior design in architecture. New York, Van Nostrand Reinhold. https://search.library.wisc.edu/catalog/999598194802121
28. Development Bureau of the Government of the Hong Kong (2011) Public open space in private developments design and management guidelines.
29. Cohen DA, Han B, Nagel CJ, Harnik P, McKenzie TL, Evenson KR, Marsh T, Williamson S, Vaughan C, Katta S (2016) The first national study of neighborhood parks: Implications for physical activity. Am J Prev Med 51(4):419–426. https://doi.org/10.1016/j.amepre.2016.03.021
30. Wan C, Shen GQ (2015) Encouraging the use of urban green space: The mediating role of attitude, perceived usefulness and perceived behavioural control. Habitat Int 50:130–139. https://doi.org/10.1016/j.habitatint.2015.08.010
31. Danford RS, Strohbach MW, Warren PS, Ryan RL (2018) Active Greening or rewilding the city: How does the intention behind small pockets of urban green affect use? Urban Forestry & Urban Greening 29:377–383. https://doi.org/10.1016/j.ufug.2017.11.014
32. Balai Kerishnan P, Maruthaveeran S, Maulan S (2020) Investigating the usability pattern and constraints of pocket parks in Kuala Lumpur. Malaysia. Urban Forestry & Urban Greening 50:126647. https://doi.org/10.1016/j.ufug.2020.126647
33. Gehl, J (2001) Life between buildings: Using public space. Danish Architectural Press.
34. Peschardt KK, Stigsdotter UK, Schipperrijn J (2016) Identifying features of pocket parks that may be related to health promoting use. Landsc Res 41(1):79–94. https://doi.org/10.1080/01426397.2014.894006
35. Chon J, Scott Shafer C (2009) Aesthetic responses to urban greenway trail environments. Landsc Res 34(1):83–104. https://doi.org/10.1080/01426390802591429
36. Rupprecht CDD, Byrne JA, Ueda H, Lo AY (2015) 'It's real, not fake like a park': Residents' perception and use of informal urban green-space in Brisbane, Australia and Sapporo, Japan. Landsc Urban Plan 143:205–218. https://doi.org/10.1016/j.landurbplan.2015.07.003
37. Dee, C (2005) Form and fabric in landscape architecture: A visual introduction. Spon Press.
38. Ma B, Hauer RJ, Xu C, Li W (2021) Visualizing evaluation model of human perceptions and characteristic indicators of landscape visual quality in urban green spaces by using nomograms. Urban Forestry & Urban Greening 65:127314. https://doi.org/10.1016/j.ufug.2021.127314
39. Marcus CC, Francis CA (Eds.) (1990) People places: Design guidelines for urban open space. Van Nostrand Reinhold.
40. Ulrich RS (1984) View Through a window may influence recovery from surgery. Science 224(4647):420–421. https://doi.org/10.1126/science.6143402
41. Whyte WH (2001) The social life of small urban spaces (8th ed. edition). Project for Public Spaces.
42. Microsoft (2005) TrueSkill^{TM} ranking system. Microsoft research. https://www.microsoft.com/en-us/research/project/trueskill-ranking-system/
43. Herbrich R, Minka T, Graepel T (2006) TrueSkill^{TM}: A Bayesian skill rating system. In Schölkopf B, Platt J, Hofmann, T (eds) Advances in neural information processing systems 19. The MIT Press, pp 569–576. https://doi.org/10.7551/mitpress/7503.003.0076
44. Salesses P, Schechtner K, Hidalgo CA (2013) The collaborative image of the city: Mapping the inequality of urban perception. PLoS ONE 8(7):e68400. https://doi.org/10.1371/journal.pone.0068400
45. Dubey A, Naik N, Parikh D, Raskar R, Hidalgo CA (2016) Deep learning the city: Quantifying urban perception at a global scale. In Leibe B, Matas J, Sebe N, Welling M (eds) Computer vision—ECCV 2016, vol 9905. Springer International Publishing, pp 196–212. https://doi.org/10.1007/978-3-319-46448-0_12

46. Beleites C, Neugebauer U, Bocklitz T, Krafft C, Popp J (2013) Sample size planning for classification models. Anal Chim Acta 760:25–33. https://doi.org/10.1016/j.aca.2012.11.007
47. Ding Y, Feng J, Ye L, Zheng S, Liu F (2024) Study on human-machine hybrid intelligent decision-making paradigm and its operational application. Computer Science 51(6):272–281
48. Settles B (2011) From theories to queries: Active learning in practice. Active learning and experimental design workshop in conjunction with AISTATS 2010, 1–18. https://proceedings.mlr.press/v16/settles11a.html
49. Fiebrink R, Cook PR, Trueman, D (2011) Human model evaluation in interactive supervised learning. Proceedings of the SIGCHI Conference on Human Factors in Computing Systems, pp. 147–156. https://doi.org/10.1145/1978942.1978965
50. Mosqueira-Rey E, Hernández-Pereira E, Alonso-Ríos D, Bobes-Bascarán J, Fernández-Leal Á (2023) Human-in-the-loop machine learning: A state of the art. Artif Intell Rev 56(4):3005–3054. https://doi.org/10.1007/s10462-022-10246-w
51. Wu X, Xiao L, Sun Y, Zhang J, Ma T, He L (2022) A survey of human-in-the-loop for machine learning. Futur Gener Comput Syst 135:364–381. https://doi.org/10.1016/j.future.2022.05.014
52. Kotsiantis S, Kanellopoulos D (2006) Discretization techniques: A recent survey. GESTS international transactions on computer science and engineering 32:47–58
53. Beijing Municipal Bureau of Statistics (2023) *2023 Beijing statistical yearbook*. China Statistics Press.
54. Ministry of Housing and Urban-Rural Development, PRC. (2024). *Guidelines for pocket park construction* ([2024]214).
55. Beijing Municipal Commission of Transport (2024) 2024 Action plan for improving the quality of Beijing's urban slow traffic system.
56. The Beijing News (2024) The forest coverage rate in Beijing is 44.9%, and the per capita park green area is 16.9 square meters. https://baijiahao.baidu.com/s?id=1799739897138398546&wfr=spider&for=pc
57. Economic Daily (2024, October 26) Creating urban oases for the people. https://www.163.com/dy/article/JFDHFU700519AKBM.html
58. Beijing Municipal Bureau of Statistics. (2022). 2022 Beijing statistical yearbook. China Statistics Press.

The manufacturer's authorised representative in the EU is Springer Nature Customer Service Centre GmbH, Europaplatz 3, 69115 Heidelberg, Germany. If you have any concerns regarding our products, please contact ProductSafety@springernature.com

Printed and bound by CPI Group (UK) Ltd, Croydon, CR0 4YY
01/10/2026
02225909-0008